Voyage to the Outer Light

Terence Thomas

PAGE PUBLISHING, INC.
Conneaut Lake, PA

First originally published by Page Publishing 2020

ISBN 978-1-6624-1389-6 (pbk)
ISBN 978-1-6624-1390-2 (digital)

Printed in the United States of America

Table of Contents

CHAPTER 1

Background

Date—2078	Subject—USS Mobius
Crew—A full complement of science personnel supervised by Science Officer Niles Barstow and Captain Felix Jenner (total crew capacity: 32)	
Mission—Deep-space probe with a primary directive of establishing robotic-operated observation bases on the outer planets	
Present location—600,000 miles from planet Saturn	
Current status—Engaged in docking procedures with the USS Stockton, an advanced class-B research station specializing in cold processing techniques	

As captain of the USS Mobius, I was delighted to hear of the success of the Stockton project because the Mobius and the Stockton were products of the same technology. Dr. Paul Stockton and I had attended the Institute for Space Exploration at the same time, so this was a reunion as well as a celebration. Back then, we shared an interest in cybernetics, and we spoke many times of the things we would do.

While I prepared for the final docking procedures, my thoughts kept wandering back to the good old times we had at the institute. The sound of the air lock quickly brought me back to reality, and as the doors slid open, I could see my old friend standing there.

I gave my customary greetings and salutations and then

introduced my science officer, Mr. Barstow. It was obvious that Paul was anxious to show us all the new things that his research had produced, but like all good hosts, he offered us refreshments.

On our way to the ship's galley, I could see the results of the work from this floating think tank. I was aware of being scanned and evaluated by the EC-1 environmental computer, and as I looked around, I was impressed by the instrumentation.

The galley was no different. It was filled with gadgets and robotic devices that were performing the most amazing tasks. We were seated and were about to resume our conversation when a young woman approached and asked what we would like. I deferred to my host's superior knowledge of the ship's cuisine, and he ordered what he described as "the greatest cup of coffee this side of Saturn's rings."

I could see that my old friend was definitely in his element here in this deep-space laboratory, and I had to admit that his unbridled enthusiasm was infectious. As we continued our conversation, we were interrupted by an announcement over the ship's intercom that the robot ship, the USS Pogo, was about to dock. We all quickly gulped down our coffee and headed to the observation deck. Despite our haste, I still caught sight of more instrumentation that made this ship the most unique ship in the fleet.

When we reached the observation deck and were seated, I looked up and was startled to see a small ship hurtling toward us on a collision course, but at the last split second, it stopped dead.

"Oh!" I yelled, "Why didn't you warn me?"

"I find that most demonstrations have more impact if you don't warn anyone first," Paul replied.

Mr. Barstow asked how the crew of the falling ship could stand the forces created by such a fast ship maneuver. "Yes, Paul," I seconded. "How can they stand the force?"

"Gentlemen, please," he replied. "I will explain, but first, come over here and observe more of the capabilities of the Pogo."

"What made you choose the name Pogo?" I asked just as the ship turned to reveal a giant robotic arm protruding from its underside.

"Does that answer your question?" Paul said laughingly.

The entire docking process was masterfully executed in just over twenty-seven seconds.

"It's as easy to park as a Porsche," Paul said as he motioned for us to follow him down the corridor.

"Captain," said Mr. Barstow. "I must report to the communications room to give them the computer interface codes."

"Very well, Mr. Barstow," I replied. "I will continue the ship's tour and meet you back on board the Mobius in one hour."

Paul and I started walking toward the cold processing lab, and he began explaining how important an element of cold was in the new technology.

"Before we spent much time in deep space, our concepts of combining substances almost always involved the application of heat somewhere along the chain of events. "When cold fusion became a practical method of producing power, we should have taken a clue from that, but another ten years passed before the first consideration of cold as a processing element was explored. Now we can produce our own fuel and food and even create the ceramic material that was used in the construction of all present-day ships.

"Our early deep-space exploration taught us that metal ships were somewhat impractical with so much onboard electromagnetic activity. Ceramic ships were the answer to deep-space exploration for they did not attract EMPs and our new computer materials were developed to be insensitive to them. Our old methods of creating gravity by spinning were clumsy and inadequate compared to the quantum-controlled gravity system, which we just witnessed."

"My god, Paul!" I said. "You haven't been wasting your time."

"There's more," he replied as we continued down the access way. When we reached the transfer corridor, we grabbed the handles as we lost the gravity of the Stockton.

I was about to ask if the new gravity system felt the same as the old system when a tremendous rush came over me.

"Wow, what was that?" I asked.

"It's the transfer from the old gravity system to the new," he said with a smile on his face.

"All external ports have scanners that identify any organic matter entering the ship and program the gravity computer for the added load. Your total weight is derived from your average density and your air displacement. We can change the weight factor manually to increase the body's efficiency, but that should not be used too often. Besides, the computer monitors for stress and makes slight adjustments automatically.

"This assures that all crew members perform at peak efficiency. The computer detects any course change and informs the gravity computer to make counterforce changes to compensate. The system is not perfect, but it enables the ship to travel three times faster without fear of centrifugal problems."

"What about inanimate objects?" I asked.

"Even better. You can adjust the system to make very heavy objects very light for easy movement. Because of the many advances, due to cold processing techniques, new materials used in the construction of ion engines have increased the efficiency by 2,000 percent. Solar collectors can produce three times the power to run this ship even this far from the sun."

Once inside the Pogo, a series of unbelievable maneuvers were initiated, and I was in for the ride of my life. Every time I could anticipate a change in direction, I would brace myself unnecessarily. I found it difficult to adjust to this new environment, and I was quick to point it out to Paul. He was equally quick to point out to me that because the system was designed to minimize the effect of unanticipated changes in a ship's momentum, acclamation to the system was unnecessary. I found the best thing was just to close my eyes.

I was pleased to know that the IGS (Internal Gravity System) was being installed in the Mobius, but I wondered why. When we docked, Paul suggested that I close my eyes, and I hardly noticed when we touched down.

Before I could tell Paul how impressed I was with the demonstration, he told me to return to my ship for six hours of sleep and then prepare for a final briefing aboard the Mobius. I was so dazzled by everything that I wondered how I could sleep, but it turned out not to be a problem. When my wake-up call came, I was ready for almost anything, or so I thought.

When I arrived at lab 2, Paul was there, drinking coffee with a somewhat ponderous look on his face. As I entered the room, Paul looked up and motioned for me to sit down. Mr. Barstow entered the room along with Medical Officer Dennison and sat on my left and right respectively. Paul stood in front of us and began by saying, "Captain Jenner, Officer Barstow, Officer Dennison, we have a problem at the outer rim. It appears that we need to investigate a phenomenon that is occurring at the very outer limits of our planetary system. A new light source has been spotted, and it appears to be moving toward us. We must know what is going on out there, and we must know soon."

I asked how soon we were scheduled to depart and was shocked when he said, "Six hours. All the details of this assignment have been programmed into your ship's main computer. What's happening out there could affect our entire system."

CHAPTER 2

The sounds of busy preparation for launch filled the air while I sat down at my control console. Systems were being checked, a course was being plotted, and many other procedures were being completed for departure. I punched in my own set of computer commands and started examining my own checklist. In the back of my mind though, I couldn't help thinking about what was waiting for us at the outer rim.

My attention was quickly returned to my screen as I observed the new procedures associated with all the new systems we acquired. I could hear Mr. Barstow calling out new instructions relating to our IGS, so I focused my thoughts to the job at hand. I announced to the crew that we were going to engage in some interesting maneuvers for the purpose of trying our new IGS system.

"Ladies and gentlemen, these maneuvers will be performed at twice the normal speed, but you will feel no inertial effects. This system compensates for many potentially dangerous situations, and in normal use, it prevents injury due to quick changes in the ship's flight path."

I further explained that it would not be necessary to brace themselves when they viewed our antics through the ship's viewports or on their computer screens.

"The effect is quite disconcerting, so if you find yourself feeling strange, just close your eyes."

I signaled to Mr. Barstow to begin launch count and then sat back to watch the response. We separated from the docking platform, and I could see everyone grabbing their consoles, their chairs, or whatever was handy. After a few unorthodox moves, the ship returned to a more conventional trajectory, and I could hear the sighs from everyone.

"Now that," I said, "is the kind of maneuvers the IGS system enables, and you should try to acclimate to the new system. We will have more drills along the way to get more accustomed to the system."

For the next few hours, our auspicious departure was the talk of the ship. I gave orders that in their off-duty hours, all crew members must take time to familiarize themselves with the new systems that were added when we were docked at the Stockton.

"We will, in fact, be testing and evaluating these systems as we make our way to Neptune. If anything unusual is observed in the operation of these systems, you should report it to Mr. Barstow."

When we were about six hours into our journey, I called a meeting of my top personnel. The meeting took place in lab 1, and everyone was on time, anxious to learn about our mission. Medical Officer Dennison spoke up first and questioned why our visit to the Stockton was so brief.

"There is," I announced to everyone, "an element of urgency to this mission. Therefore, we must learn on the run, so to speak. I am sure that you all have questions, but I must tell you that we know very little about what we are approaching."

"What exactly are we looking for?" came a voice from the back of the room.

"Light," I replied.

"Excuse me?" again came a voice from the back of the room. "An unexpected source of light has appeared at the outer rim of our system. We have been assigned the duty of investigating this phenomenon to determine the source and possible effects it may have on our system. Our IGS will enable us to get there in record time to get an evaluation as soon as possible. Another point of concern is that all spectrograph analysis charts of the phenomenon have produced strange results. Because of unknown elements, we face a substantial risk, so I expect everyone to be alert. Are there any questions?"

"Do we have any theories about the phenomenon?" asked one of the medical staff.

"We haven't had enough time to form any substantial theories, but since its first appearance, we have constantly monitored it. If there are no further questions, this meeting is adjourned."

For the next few days, a feeling of excitement and anticipation filled the air. Everyone seemed anxious to get on with the job, and theories were popping up everywhere. The information we were compiling continued to be confusing, but my crew was determined to find answers. Systems diagnostics were being run to assure the accuracy of our analysis. Enigmatic results persisted. However, the spirit of the science crew remained high. I was sure that this positive attitude could be attributed to all the newly installed systems that were functioning so impressively.

I took a stroll down to the infirmary to check on things, and I was met by Chief Medical Officer Dennison.

"I have a few things I'd like to discuss with you," she said.

"Then by all means," I replied.

"In the last twenty-four hours, I have had nine patients suffering from dizziness."

"And what do you think is the cause?"

"I think there may be some causal effects related to the IGS unit."

"Can you say that for sure?"

"No."

"Then keep working on it and keep me informed."

I left the lab and started walking to the bridge when all of a sudden, I felt dizzy. It passed in a minute, and I continued down the corridor. When I arrived at the bridge, I was greeted by Mr. Barstow.

"What's our status, Mr. Barstow?" I asked.

"We are on course and schedule."

"And our destination?"

"Still inconclusive, sir."

"Carry on, Mr. Barstow. I'll be in my quarters if you need me."

I walked down the corridor with some difficulty and collapsed on my bed. As I lay there, my mind was spinning with thoughts of the last twenty-four hours. I was tossing and turning, restless for no reason or no reason I could identify. Something was wrong, very wrong, but I didn't know what. I finally fell asleep, if you could call it sleep, and what seemed just minutes later, I received a wake-up call. Drowsy and feeling like something nasty warmed over, I stepped to my console and called up my schedule.

I had a protocol meeting with five new members of our crew that we picked up from the Stockton. I had just a few minutes to get ready for my meeting, so I stepped into my lavatory and splashed some water on my face. Although I felt better, I was not quite my joke-cracking self. But duty was calling, and the conference-room door buzzer was buzzing.

"Reporting for protocol meeting, sir." He paused. "Ensign Clark, sir."

"Yes, come in, ladies and gentlemen."

"I know that I speak for all of us when I say that we are honored to serve aboard the Mobius, the most advanced science ship in the fleet."

"And the only class-D ship in existence," I added proudly. "Have a seat, everyone. Can I get you something?"

"No, sir. We're anxious to get on with it."

"Well then, let's do so. Sit down, everyone."

"The ship is so big and luxurious," said one young lady. "And you are?" I asked.

"Oh, I'm Mineralogist Jan Cooper, sir."

"Ms. Cooper, we try to make the ship as comfortable as possible, and I assure you that some parts of the ship would not be described as luxurious. Since our missions last for a considerable length of time, we try to have as many comforts as possible. We have a handball court and even a holographic golf course. Our hydroponics lab keeps us replete with vegetables, and our food replicators can make almost anything from vegetables. Vitamin supplements are added to our food to ensure our health to keep us alert and function well. We also have a medical lab that can take care of anything that a major hospital could handle. I see that one of you is an assistant medical officer."

"Yes, sir. Diana Kane," she replied.

"Well, Ms. Kane, you come to us with quite a list of credentials. You worked on the Stockton in the research department with Dr. Morgan and developed a new strain of bacteria that helps maintain blood balance."

"Yes, sir."

"Well, we are certainly glad to have you aboard, Officer Kane. I see we have twins, Lorin and Lanie Rodgers."

"Yes, sir," both responded together.

"It says on my report that you are both chemists."

"That's correct," they said again in unison.

"And of course, last but not least, Ensign Clark. I see you have served with distinction on the USS Paxton. You saved two crewmen's lives in an explosion. Quite remarkable. All of you…welcome aboard the Mobius. I have some business to take care of just now, so let's meet on the bridge in one hour."

CHAPTER 3

As I stepped out of the elevator, I could see Science Officer Barstow talking to the chief engineer.

"Mr. Barstow," I said with authority.

"Yes, sir," he replied confidently.

"How are things going?"

"We still can't identify the light source, but I'm trying something new that we had installed at the Stockton. We might as well make use of the new equipment."

"I would like a briefing on our progress when the new equipment has been tested."

"I'm not certain when that will be, Captain, but I will keep you informed."

"Carry on, Mr. Barstow."

"Yes, sir."

I left the bridge with a certain amount of optimism for I had the best science officer in the fleet, and if anyone could solve our problems, it would be Barstow. Still I had an ominous feeling with the mission, the kind of feeling that raises the hair on the back of your neck. I tried to keep my mind on the work at hand by concentrating on the very active schedule on learning the new system that we had installed, but all I really wanted to do was go to sleep. I still had my protocol meeting, and I was not up for it. I stopped for a moment to steady myself and stumbled to the floor.

"Captain Jenner, Captain Jenner!" I could hear someone speaking as I came to. "Wake up, sir! Please wake up!"

"I'm all right. I must have blacked out."

"Well, we are going down to the infirmary."

"I can make it on my own."

"Nonsense. I was just on my way to sick bay, sir."

"Well, in that case, let's proceed."

"What happened, sir?"

"I'm not sure. I was walking, and then…"

"Captain, what happened?" I heard and looked up to see Officer Dennison approaching.

"I was just explaining to Officer Kane. Let's wait until we get to sick bay."

I could see the doorway just ahead, and as we entered, I was relieved to lie down on the examination table. Officer Dennison proceeded to the medical analysis console and punched in my code. Scanning started, and in a few minutes, it was complete.

"I can't find anything wrong with you, Captain," she said, apparently puzzled at her findings.

"In our briefing at the beginning of the mission, it was mentioned that our gravitational environment included adjustment for our peak physical performance. Could this be out of adjustment?" I said.

"That could be possible," said Officer Dennison. "Diana, run a diagnostic on the physical enhancement circuits of the IGS." She immediately started the diagnostic programming, and in just a few minutes, she found an error.

"That's extraordinary," said Ms. Kane.

"Yes it is." echoed officer Dennison. "Why would you think of that, Captain?"

"Because I have never felt anything quite that strange. It just didn't feel right. Arrange for all crew members to be scanned in the next twenty-four hours, Diana."

"Yes, sir," she replied.

"Did you find something else?" asked Dennison.

"I believe so."

"Well, I will leave you two to work out the details," I said. "I'll be on the bridge."

"Are you feeling okay, Captain?" said Diana.

"Yes, I'm fine. Now I must go. Oh, Officer Kane, I'll see you on the bridge in twenty minutes for our protocol meeting."

"Yes, sir," she said in a self-assured manner. It must have been the electrolytes, I thought to myself.

The first order of business when I arrived on the bridge was to inform everyone of their examination within the next twenty-four hours. "Officer Barstow, I want a running diagnostic of all the new systems on this ship."

"Already done, sir," Barstow replied. "How's that?"

"I spotted the diagnostic that was running in sick bay, so I decided to run a continual check on all systems with the Ferret 1."

"And what is Ferret 1?"

"It's a system of my own design."

"And why was I not informed?"

"You have enough to take care of running the ship, sir. Since I am the head science officer, I felt the Ferret 1 might be useful."

"Well, I want a full briefing on the Ferret 1 at our next scheduled conference."

"Yes, sir," he said confidently.

I turned and walked away smiling to myself. I've got the best crew in the fleet.

CHAPTER 4

While leaving the bridge, I saw the new crew members heading my way.

"You're here for your protocol meeting."

"Yes, sir," said Mr. Clark.

"Have you become the spokesman for the group, Mr. Clark?"

"Ugh, no, sir. I…"

"Never mind, Mr. Clark. Let's get to it. As you can see, every twenty feet are our communication modules, in case your personal communicators do not operate. The ship has seven decks, three labs, a recreation area, and two maintenance robots. And we have recently, as you know, been fitted with the latest in gravitational and communicational technology. The ship was constructed in space from a new ceramic material that is insensitive to electromagnetic radiation. Let's proceed to lab 1."

As we headed down the corridor, I could tell that they were impressed with the ship. We stepped into the elevator, and I pressed the button for level 2. "In addition to this elevator, we have another, which just provides access to three labs. Our many stairways are there to give us the opportunity to get required exercise." The lab doors slid open, and we entered.

"To the left, you can see one of our three electron microscopes, one in each lab. The next section is our mineralogy lab, which should be of interest to you, Officer Cooper. It comes with the new SP-900 spectral analyzer and LASER sample processor. Samples are gathered from all four of our frontal probes and can be analyzed from a distance or transported down the tube to a direct access terminal. Our next station is the chemical lab, complete with isolation chamber, hand manipulator, in an anti-contamination access chamber." I stepped back a moment to see the reaction to all the instrumentation.

"This is quite incredible," said Mr. Clark.

"Oh yes, Mr. Clark. I have a surprise for you also," I said. "You will be working with our communications officer Ashley Yates on one of the most advanced interfacing computers that technology has to offer."

"You mean the C-400?"

"Yes, Mr. Clark. The C-400."

"Wow, sir. That's great."

"Control your enthusiasm."

"Yes, sir," he said, embarrassed.

"Next, we have our DNA station manned by our electronics officer, Max Palasco, and our botanist, Officer Richard Greyson. He grows plants in our hydroponics garden and is responsible for the plants you see around the ship. Are there any questions?"

"Yes, sir," said Ms. Cooper. "I am not familiar with the retrieval system."

"Just hit any touch pad on your terminal, and you will activate a full tutorial on all new systems that you will need. This applies to everyone. Are there any other questions?" I looked around, and there was no response. "If you think of any questions, just ask me or Science Officer Barstow."

I cut the meeting short because I was feeling dizzy again. What was going on with this? I quickly made my way to my quarters and sat down on my bed. This was beginning to worry me.

CHAPTER 5

The next few weeks were uneventful and disappointing. We were no closer to finding any answers, and I could see discouraged looks on the faces around me. I tried to encourage the crew to keep busy and to keep a positive attitude. We did have new systems to learn and our regular duties to attend. We were also learning about one another. I found out that Mr. Clark was quite a musician and was a member of the symphony orchestra at the academy.

Mineralogist Cooper was a painter and had several gallery showings. Chemists Lorin and Lanie Rodgers were table tennis enthusiasts. What better activity for twins? Engineer Seth Langer was an Olympic skier, and Communications Officer Ashley Yates, like myself, loved to play handball.

We were actually beginning to make progress in identifying the light source although this was more a process of elimination. Systems were operating well, and the new capability of the equipment was exciting to everyone. My dizzy spells had disappeared, and we seemed to have everything under control. The crew was getting back on track, and morale was definitely getting better. I felt confident that we were going to continue this positive trend. I scheduled many update meetings to keep the enthusiasm up. The banter was stimulating, and everyone was getting more involved in this mission.

I initiated a series of tests for all our new systems, and this also seemed to stimulate everyone's imagination. I even ordered a test of our propulsion and gravitational systems. I gave the order on one test, and everything checked out for about twenty minutes. And then alarms started going off all over the ship.

My immediate response was to order a power down. But the ship's controls failed to operate properly, and a kind of chaos took over. Our computer screens filled with information indicating all equipment failures as tension filled the air.

For a few moments, we were on the verge of panic, and then we suddenly regained control. Everyone stood silent for a moment, and we all looked at one another, as if asking for an explanation. I took a deep breath and yelled, "Damage report, engineering!"

"It's Matson, sir," came a voice over the intercom. "Captain," he continued, "we have incurred no damage, and we are trying to determine why we lost control."

"Run a full diagnostic analysis on all navigational systems, and do it on the double!" I said and motioned to Barstow to follow me to lab 2.

While walking, I tried to get a sense of what Barstow was thinking. However, he seemed reluctant to voice out his opinion. It was common knowledge that science officers did not like to speculate. Barstow, on the other hand, was never shy about expounding his theories. I was surprised at his silence and concerned about the look I saw on his face.

As soon as we entered the lab, I pushed the intercom button and asked the chief engineer to put the diagnostics information on screen when it became available.

"We'll be right down," I announced to the chief engineer. "Mr. Barstow, we are headed to engineering."

"Yes, sir," he said with a reassuring confidence.

We walked quickly with purpose and determination. Neither of us said anything, but you could almost hear the gears grinding.

"Officer Matson," I bellowed upon arrival. "What's your assessment?"

"Sir, we have registered fourteen hull breaches."

"Hull breaches from what?"

"I'm not certain, sir."

"You mean that the ship's integrity was compromised fourteen times and you don't know what hit us?"

"Sir, we didn't register impact of any kind, and we haven't lost pressure."

"Well, check all systems, Mr. Matson. Mr. Barstow, what's on your mind?"

"Only theory, sir."

"We need anything we can get."

"I can't speculate about the hull breaches, but I believe the source of light is much closer than we think. Too close to be a star."

"What do you think it is?"

"I can't say for sure."

"And you don't want to speculate."

"No, sir. I have narrowed down the possibilities, and I would like to schedule an update meeting to review our findings."

"Set up a meeting in three hours. If you need me, in the meantime, I will be on the handball court." I went to my quarters and changed into my sweats. While walking to the court, I could feel renewed energy, and I was anxious to get started.

I imagined the ball was the nose of everyone I disliked, and I had quite a game going. Anger was a great motivator. With each hit, I could feel the release of tension. When I rested for a moment, I could see that someone was joining me on the court. It was Communications Officer Ashley Yates.

"Care for a game?" she said.

"Certainly. You serve first."

"What do you think it is, Captain?"

"What do I think what is?"

"The light."

"I'm not a scientist."

"You must have an opinion."

"I don't care to speculate." I slapped the ball hard, and she began her conversation again.

"Why don't we run diagnostics?"

"We have."

"I mean continually, using the secondary backup comparison mode."

"Your serve, Officer Yates."

She served the ball hard, but I managed to return it.

"The new systems use less power, so the added load would be insignificant."

"You seem to have done your homework, Ms. Yates."

"Yes, sir."

"Then you will be in charge of setting it up, but check with Officer Barstow."

"Yes, sir," she said with unbridled enthusiasm.

"But first, let us finish the game."

"Of course, sir," she said with a grin on her face.

As we played, I noticed for the first time how attractive my communications officer was. She was a perfect match for one of our engineers. What was I thinking? Am I becoming a matchmaker on my own ship? Where were these thoughts coming from?

CHAPTER 6

Our workout was exhilarating. It was such a pleasure to be with that vital young woman. She reminded me of an old flame at the academy. We were in quantum physics class together, and she was always coming up with new theories to explain everything. Her energy could sweep you away, and her enthusiasm was infectious. She was the love of my life, and when I heard that she was killed on a routine training mission, I was devastated. I was a basket case for a while, but I threw myself into my work and became the youngest commissioned captain in the fleet. I will never forget her for she occupies a special place in my heart and mind. That night, I fell asleep thinking of our good times at the academy.

For the first time in weeks, sleep came easy. A montage of pleasant images ran through my mind. Visions of the happiest moments of my life flashed before my eyes, and though my sleep time seemed too short, I woke up refreshed.

My first order of business was to check with Chief Engineer Matson. "Chief Matson," I announced to engineering. "I'll be down in one minute."

"Yes, Captain," replied Matson with the sound of confidence.

I was glad to hear that snap in his voice. He had never failed me, and I was sure he had some important answers. On my way to engineering, I passed Medical Officer Kane.

"Captain Jenner," she said. "I would like to have a word with you when you have the time."

"I'm on my way to engineering, so when I finish there, I will stop by lab 2."

"That will be just fine, sir."

I continued down the corridor at a fast clip, and soon I found myself at the door of engineering. The door opened, and I looked around. "Mr. Matson," I bellowed with authority. "Do you have answers for me?"

"Yes, sir, I do."

"Why did we have fourteen false hull breaches registered?"

"There was an error in sensor programming that got overlooked while we were installing the new systems, sir."

"Has it been corrected?"

"Yes, sir. I'm happy to say it has, and we have everything reset and ready to go."

"I would like a full report on the incident on computer before you leave engineering."

"It will be done, sir."

"See that it is."

I could see he was a bit shaken by my stern attitude, but he was a seasoned officer and could take the pressure. He had been there before, and he always performed exceptionally well under stressful situations. And I didn't expect this to be any different.

I was curious about what Officer Kane had to say, so I cut short my meeting with Matson. I was also lucky to get Kane for this mission. She was well-known around the fleet, and she was in demand. Her work on the Stockton earned her the highest medical award given at the academy, the Gold Medallion. She was in the effect of the IGS on body chemistry, and that was why she signed on. When I arrived at lab 2, she was busy at the analysis console. I stood there watching while she called up information on the computer. Her intensity was evident, and when the information appeared on screen, she reacted almost gleefully.

"Having a good day, Officer Kane?" I asked.

"Oh yes, sir. I mean, yes, sir."

"What do you have to report to me?"

"Well, sir, when I was aboard the Stockton, I theorized that the

IGS might have an effect on body chemistry and could cause problems."

"Did you use the Pogo for your preliminary tests?"

"Very good, sir. Yes, I did."

"And what were your findings? What do you attribute that to?"

"A number of things. First, the system aboard the Pogo was not as powerful as the one here. Secondly, the crew aboard the Pogo lived aboard the Stockton and had only sporadic exposure to the system."

"And now that we have all been exposed to the system constantly for weeks, what have you discovered?"

"As I suspected, a slight chemical imbalance exists in all of us that can create a variety of symptoms including dizziness."

"Do you think that that may be the cause of my dizziness?"

"Yes, sir, I do."

"What do you prescribe, Officer Kane?"

"It's a synthetic compound that I have put together here in the lab."

"How is it taken?"

"Orally."

"Great. I hate needles."

"We don't use needles on starships."

"Not even when requested?"

"If you are looking for pain, I could punch you after you take your medicine."

"That won't be necessary."

"And they say that starship captains don't have a sense of humor."

"Who spread that ugly rumor? Probably the same person who said that medical officers don't have a sense of humor."

"Well, maybe we're special, Captain."

"I've always known that I was."

"Why is that?"

"I can dance."

"Oh, really, Captain?"

"Yes, I live to dance."

"So what are you doing on a starship?"

"I just do this part-time, between engagements."

"Engagements?"

"As a tap dancer. I perform at the academy."

"Ha ha, I would like to see that."

"I'll get you tickets to see my next performance."

"I'm counting on it. Now here's your dose. Drink it down."

"It's green. Doesn't look very appetizing."

"Oh, drink it down, you big baby."

"Right."

"Drink it all."

"You know, this is not bad."

"Good. Now get out of here so I can prepare the doses for the other crew members."

"Very well, Officer Kane. I shall return to my duties."

CHAPTER 7

It was enjoyable to banter with Officer Kane, but it was time to get back to duties. Mr. Barstow was waiting for me in conference room 1, and as usual, he was there on time.

"Mr. Barstow, I'm in a good mood, and I'd like to stay that way."

"I think you'll be happy with our progress, Captain. Well, sir, first observations gave it the light intensity of a nova. But it does not have the density of a nova, and novas do not just appear out of nowhere. Its density is not even sufficient to classify it as a planet. What it is cannot be determined by a spectral analysis, which means that its substance is outside our ability to identify. I do feel that it is much closer than previous estimates."

"How much closer?"

"Somewhere near Neptune. We'll know more when we get closer."

"How long will it take, Mr. Barstow?"

"Approximately one month."

"Approximately?" I questioned.

"Actually, twenty-eight days and eleven hours."

"Now that's what I like to hear. Carry on, Mr. Barstow."

"Captain, the information you requested about the ferret system is on your computer terminal. Just press F and hold it down and then press one. The system will appear on your screen in about four seconds. Actually, 3.925 seconds."

"Thank you, Mr. Barstow," I said with a smile on my face.

I could tell that Barstow was concerned about our mission and he was troubled by his inconclusive report. He would work on this problem relentlessly, and I had the utmost confidence in his ability. He had worked at the academy and was responsible for research that made major changes in both techniques and instrumentation of our starships. One of our mission objectives was to perform an experiment that was conceived by Barstow.

When I was commissioned to serve as captain aboard the Mobius, Barstow was my first choice. He brought with him my chief engineer Matson, robotics engineer Alice Grason, and maintenance officer Cal Decker.

Medical Officer Lisa Dennison had been the only chief medical officer the Mobius had ever known. Botanist Richard Greyson was transferred from the now-decommissioned USS Utah. Max Palasco and Kile Basser, both electronics engineers, were the top of their class graduates of the academy and requested this assignment specifically. Biron Blake, a very creative maintenance engineer, was a friend of Cal Decker's and came highly recommended. Lee Brock and Mark Anderson were navigation engineers and had been with the Mobius since its maiden voyage. My communications officer Ashley Yates became a crew member on the second mission of the Mobius and worked with Alice Grason on robotics in her spare time.

Although the ship could accommodate thirty-two, we had never had more than twenty-two crew members for a mission. The nineteen crew members we had were more to my liking. It gave everyone more breathing room.

As a kid, being a starship captain was the only thing I ever wanted to be. My family was not too enthusiastic about my career choice, but I would not be denied. I just couldn't be happy as a businessman like my father.

I was a bit rebellious in school and did some things that I'd rather forget. I did graduate with honors though and enrolled at the academy. My first duty was aboard the USS Constellation, the largest deep-space ship in the fleet. Being first officer, I was really impressed with myself, but soon I found out what responsibility was all about. Under Captain Ronald J. Lassiter, I learned some hard lessons and developed a taste for adventure.

After serving for seven years on the Constellation, I was given my own ship, the USS Apache, a service-and-supply ship with a crew of ten. It may not have been much compared to the Constellation, but it was my ship. Five years later, I was given the opportunity to captain the Mobius, so I signed on. In the years that I have been captain of the Mobius, we had had seven missions, but none more important than this, which was why I had such a special handpicked crew.

Reports on the light source that we were investigating said that it appeared out of nowhere as a tiny dot that increased in brightness by 50 percent in just three months. Its actual distance from Earth could not be determined, and spectral analysis yielded strange results. Although novas are stars that increase in brightness over a period of time, they have a spectral signature that in no way resembled the results obtained from the new light source. The only conclusion that could be derived was that the light was coming from a gigantic mass of amorphous material. Nothing this size had ever remained unidentified for such a long time, giving rise to much concern about its effect on our system.

I was not happy about having to adapt to all the newly installed systems when we should be concentrating on the task at hand. It was true that the IGS would shorten our travel time by more than six weeks. However, the annoying glitches were disconcerting.

I stopped by the medical lab to see how everyone was getting along. Medical Officer Dennison reported that other than a few cuts and bruises, things were kind of dull.

"That's what I like to hear," I said with a smile on my face. "I want this mission to bore you to tears." I continued, "What about you, Ms. Cooper? Have you discovered anything interesting in your collection probes?"

"Just normal cosmic debris. Nothing to get excited about," she said without looking up from her console.

"Well, it's good to see that everything is under control," I said with a touch of sarcasm.

"Captain," said Ms. Kane, "it's great to know you're so concerned about how stimulating our work is."

I smiled and walked away. It was great to see that my crew was comfortable enough to give me a hard time. A sense of humor was an important thing to have on a starship.

CHAPTER 8

For two weeks, the ship was bustling with activity. My talented crew was living up to its reputation. Our robots were fitted with new devices that made them capable of dealing with a wider variety of situations. Medical breakthroughs were happening on a daily basis. Mr. Barstow developed light-sensing devices that were twenty times more sensitive than anything we were using. Officer Kane had developed a nutrient that stimulated the electrolyte production in the body, and it actually tasted good. Our botanist, Richard Greyson, and our two chemists, Lorin and Lanie Rodgers—in a collaborative effort—discovered a chemical additive that increased the efficiency of the oxygen production in plants. I, too, not to be outdone, shaved three strokes off my golf score. Everything was going so well I decided to schedule a little celebration.

We all gathered in the main conference room for some wine and conversation.

"May I have your attention please," I said loudly. "We are gathered here to acknowledge some of the achievements of our colleagues. We've had some times now to get to know one another, and I can say for myself that I am proud to serve as captain of such an exemplary group of talented young scientists. And to show how much I respect you all, this speech is over. Let's all have a glass of wine."

They all laughed, and we made our way to the bar. "Are you tending the bar, Mr. Clark?" I asked.

"Yes, sir. If it's all right with you, sir," he said nervously.

"Of course, it is, young man. I'll have a scotch on the rocks."

"Rocks, sir?" He was puzzled.

"Scotch whisky with ice."

"Oh, of course, sir," he said, smiling.

He made the drink, fumbling for everything but finally succeeding. I turned and saw the Rodgers twins approaching the bar.

"Ladies, can I order something for you?"

"White wine," one of them said. I was not sure which one.

"I hear that you have combined efforts with our botanist, Mr. Greyson."

"Yes, sir, we did," once again, one spoke.

"When I get some time, I would like to talk to you about your achievements."

"Thank you, sir," they said in unison.

"Mr. Clark, two white wines for the ladies." I saw officers Kane and Dennison waving across the room. "If you will excuse me, ladies, I must have a word with my medical officers." The twins nodded, and I walked over to the pair of obviously giddy medical officers.

"How do you tell them apart?" said Ms. Dennison laughingly. "If I avoid eye contact, then they don't know which one I'm speaking to."

"I knew he'd find a way to handle that," said Ms. Kane to Ms. Dennison.

"You ladies seem to be having a good time."

"Yes, we are," they said surprisingly in unison.

"My god, we're beginning to sound like the Rodgers twins," said Ms. Dennison, again laughing.

"I think you ladies should find a table and sit down. I'll join you in a moment."

I could see botanist Richard Greyson across the room, and I walked over to him.

"Mr. Greyson, I hear you have been collaborating with the Rodgers twins."

"Yes, sir. And we have come up with some interesting results."

"An increased efficiency of 7 percent in oxygen production in our plants, I hear."

"Actually, 9 percent, but who's counting?"

"Mr. Greyson, I do believe I can count on you."

"Yes, sir."

"Oh, Ms. Cooper. Have you met Mr. Greyson?" I said as she walked by.

"No, I haven't," she replied.

"Mr. Greyson, this is our mineralogist, Ms. Jan Cooper. I'll leave you two alone to get acquainted."

I spied my chief navigator at the door looking around. I assumed for me, so I made my way to him. He saw me coming and waved me over.

"Is there something you need to speak to me about, Mr. Anderson?"

"Yes, sir, if we may have a word on the bridge."

"Certainly," I replied in my most official voice.

We walked briskly and said nothing. When we stepped onto the bridge, I asked what was on his mind.

"Sir, it's the light source," he said nervously. "Yes, what about it?"

"Well, sir, it stopped moving."

"Is that possible?"

"It would seem so, Captain."

"You're sure about this?"

"Yes, sir. I can't explain it."

"Have you spoken to Barstow about this?"

"Yes, sir."

"And where is Barstow now?"

"He is at his personal computer terminal."

"Get him up here on the double."

I went to the conference room and poured a cup of coffee. I sat down, and by my second sip, Barstow appeared.

"You wanted me, sir?"

"Why didn't you inform me?"

"You were celebrating, sir. And it's not an emergency, so I didn't think it was necessary."

"Then why did Mr. Anderson summon me?"

"That's my fault, sir," said Anderson. "I just panicked."

"Mr. Barstow, we are going to a higher alert level. I will return to the party, but in two hours, I'll be back to discuss the details of anything you have discovered about this new development."

When I returned to the party, everyone was getting along just dandy. The ladies were still at the table where I left them.

"Ladies, would you tell me what transpired in my absence?"

"Well, you missed a big brawl," said Ms. Dennison.

"Oh no. I missed a good fight."

"No, we're lying."

CHAPTER 9

I went back to the bridge and saw Mark Anderson sitting at his post. "All right, Mr. Anderson, let's get down to it," I said with authority.

"I think we should wait for Officer Barstow, sir."

"Very well."

"Oh, here he comes now," said Mark, as if relieved. "Mr. Barstow, what have we got here?"

"What we have is an amorphous mass that is located near Neptune, yet it is unaffected by Neptune's gravitational pull."

"What could account for this?"

"Some kind of force."

"What kind of force?"

"Intelligent force."

"Something is controlling the force?"

"It would appear so."

"From where?" I asked. "Undetermined."

"Wouldn't we be able to detect any control power source?"

"Not necessarily. Power beams emanating from unfamiliar technology could remain undetected if we don't know what to look for."

"Even beams powerful enough to control such a large mass?" I asked.

"If modulated or masked."

"Mr. Anderson," I said as I turned to him. "You've been very quiet. What do you have to say?"

"Well, sir, I've never have I seen anything like it. A huge mass changing its momentum. It just doesn't happen."

"Was anything visible?"

"Even beams powerful enough to control such a large mass?" I asked.

"If modulated or masked."

"Mr. Anderson," I said as I turned to him. "You've been very quiet. What do you have to say?"

"Well, sir, I've never have I seen anything like it. A huge mass changing its momentum. It just doesn't happen."

"Was anything visible?"

"No, sir, but the spectral shift detectors were unmistakable and even backup systems gave the same readings."

"Mr. Barstow, have we made any progress on our spectral analysis?"

"The results are more confusing the closer we get."

"Could the signals Mr. Anderson received be wrong since we have these spectral problems?"

"Not likely," said Barstow. "Our spectral analysis gives us no results at all. Mr. Anderson's results were from instruments that were designed only to detect color shifts indicating a change in direction for navigational purposes, and because a change was observed on multiple systems, it is not likely to be a false signal."

"Well, Mr. Barstow, please notify me the moment you can nail down this spectral problem. And I mean the very moment. Even if I'm asleep, wake me. We need some answers."

"Yes, sir. I will and I may need to make some alteration in our spectral analyzer."

"Changes, Mr. Barstow. Do you have some theory?"

"Nothing I care to speculate about just now."

"Get to it, Mr. Barstow. I want some answers soon. And, Mr. Anderson, you are off duty. Get some sleep."

"Yes, sir. I'm sorry I interrupted the party."

"Don't give it a second thought, Mr. Anderson." I then addressed the both of them, "I'm counting on the both of you to come up with something soon. I know you can do it. Now I'm due some sleep myself, so I'll see you when you are back on duty."

For the first time in weeks, sleep did not come easy. Not being able to identify the object of our destination was getting on my last nerve. I was trying to keep my mood to myself, but I was sure that everyone could tell I was stressed. Everyone had their work to keep them busy, so I spent time on the golf course. My score was terrible and getting worse.

Mineralogist Cooper summoned me to lab 2, and it felt good to be involved. When I got to the lab, she was working frantically at her station.

"Captain, I must see you about something."

"Yes, Ms. Cooper, how can I be of service?"

"There is something wrong with our collectors."

"What is the problem?"

"I collected some yellow powder and was preparing a slide in the isolation chamber, and the powder disappeared. I checked for drafts or any air circulating in the chamber, and no air was circulating."

"Have you spoken to Mr. Barstow about this?"

"No, sir."

"I think he should take a look. Just punch in 21 and L2. How much powder was there?"

"Not much. But I had a couple of slides, and both disappeared."

Barstow entered the lab and came over to us. "Mr. Barstow, we have a problem. Ms. Cooper was preparing slides of a powder from our collection tubes, and it disappeared in the isolation chamber."

"We do have several access ports that decontaminate the sample chamber and an access count on the computer," said Mr. Barstow. "Because of the importance of containment, there are four hundred sensors that can detect even a single quark."

"Thank you for that information, Mr. Barstow," I said. "Would you please check the access count and find out if we have any foreign matter in the chamber?"

He punched in the codes, and the information appeared on the screen—zero access count and a completely empty isolation chamber. A red light alert started flashing.

"What's that, Mr. Barstow?" asked Ms. Cooper.

"That's strange. The internal volume is incorrect. I will try to recalibrate."

He entered more codes and still ended up with the same results. After still another try, he failed again. "Sir, I must see to this problem immediately."

"Ms. Cooper, I want you to take a break until further notice."

"But, sir—"

"Take a break, Ms. Cooper. That's an order. I will leave you to your work, Mr. Barstow. I'll be on the bridge."

He was too busy to respond, so I stepped quickly and soon was standing behind Mr. Anderson. "What is that, Mr. Anderson?" I asked while looking at something amazing on the large screen.

"It's our light source, Captain," he said.

"My god, it looks like a nebula."

I pressed the intercom button and asked Mr. Barstow to report to the bridge immediately. He arrived in record time and soon was looking at the incredible sight. We all stood there gazing at the most incredible sight we had ever seen.

CHAPTER 10

I had a lot on my mind as I entered my quarters, and when I stepped to my bed, I was grabbed from behind and wrestled to the floor. I could not believe what was happening. I yelled out, "What are you doing?"

"Shut up!" I heard someone say.

"Grab his legs!"

I tried to fight back, but there were too many of them.

"What do you think you're doing?" I yelled.

"Shut him up," I heard someone say, and someone stuffed a rag in my mouth. They had me pinned down and were working me over when one of them pulled out a knife and held it to my throat.

"I want to kill him now," he said, pressing the blade against my skin.

"Not yet. I have a few more kicks to get in."

He continued to kick me in the stomach. I wondered how long I could survive this torture.

"I'm going to kill him now," said the man with the knife. "Go ahead. Kill him."

He slid the knife quickly across my throat. I lay there watching the blood pour out on the floor and wondered why I was still alive. Suddenly, I heard the ship's alarm, and when I looked down, there was no blood and no men in the room. It was the most vivid and horrifying dream I ever had. My mind was swimming in a sea of imagery. I staggered about the room as in a drunken stupor. My mind cleared, and things seemed normal except for the ship's alarm. That was real.

I dressed quickly and made my way to the bridge. Mr. Barstow was giving orders.

"What's the matter, Mr. Barstow?" I asked excitedly.

"Hull breach," he replied.

"If we have a hull breach, how come the console has no indication of pressure loss?"

"We are attempting to find the answer to that question now, Captain," he answered with his customary confidence.

I stepped to the console and made an announcement. "We are initiating Procedure 19." I motioned for Barstow to hand out the packs, and for the next few minutes, each crew member was given a pack. The packs were test instruments that could test any instrument on the ship. Just pull out the chord and plug it into any jack around the ship. They were located at every terminal. Insert the jack and press the red button; and it will repair, and it would repair any defects on the spot. The pack would also map out the system for you and display them on its screen.

Each crew member was assigned an area to check, and when they were finished, they were to report back to the bridge.

"What do you think Procedure 19 will reveal, Mr. Barstow?"

"That there was no malfunction in our main system."

"Why do you feel that way?"

"We had no loss of pressure."

"Then what triggered the alarm?"

"Something we are overlooking."

"Then look it over, Mr. Barstow."

"Yes, sir."

I watched my science officer leave the bridge and wondered if he was as frustrated as I was.

"Sir," said my navigator. "Yes, Mr. Anderson."

"A camera scan of the outside of the ship has revealed a yellow substance all over the ship."

"Get Alice Grason to the bridge immediately."

He punched in the paging code and looked up for further

instructions. "Why have our sensors not detected this? Scan to the left."

"It's everywhere," he observed."

"You paged me, Captain?" I turned to see Ms. Grason standing there.

"I want you to get Plato and Archimedes ready for an outside mission. How long will it take?"

"Just a few minutes will be enough time to load them into the air lock."

"Go do it and report back here when you're finished."

"Yes, sir, I'm on my way."

"Keep focused right where you are, Mr. Anderson. The robot access ports are just to the right, and we should see them in just a few minutes." We both remained silent as we looked out at the top of the ship. Minutes seemed endless as we viewed the large screen; and then two robots appeared on screen, and Ms. Grason appeared at the doorway.

"Aren't you controlling the robots, Ms. Grason?"

"No, sir. The new programs allow the robots to act independently until it encounters anything unexpected, and then it waits for further instructions. That way, they are not forced to follow programming that might create danger situations. The programs are far more accurate than manual control, but manual override is easily engaged."

"I'm very impressed, Ms. Grason. Those new programs, are they your designs?"

"Yes, sir. They are."

"Very good, young lady. Explain what they are programmed to do now."

"If you'll look at Plato…"

"Which one is Plato?" asked Mr. Anderson.

"It's the one with Plato written on his side," said Ms. Grason. "I'm sorry," Mr. Anderson said with an embarrassed blush on his face.

"Plato has turned on the minicam attached to his arm. We'll get a better view if we transfer to robo-cam," she continued.

"Make it so, Mr. Anderson," I ordered.

He punched in the proper code, and the screen changed to a view of the robot sample chamber as it collected a sample of the yellow powder.

"Continue, Ms. Grason," I said.

Since this is the only instruction that Plato was given, he'll return to the air lock immediately. He will then disconnect the sample chamber and place it in a transfer port to be transported to an examination chamber."

CHAPTER 11

I asked Mr. Anderson to page Officer Cooper, and he reminded me that she was involved in a Procedure 19 test.

"Tell her to discontinue the test and report to the bridge on the double," I ordered.

"Coffee, Ms. Grason?" I asked.

"Thank you, sir."

"How about you, Mr. Anderson?

"Not just now, sir."

I stepped to the coffee module and asked her how she liked it.

"Black. No sugar."

"Just the way I like it. Now, Ms. Grason, I was very impressed by your demonstration, and I feel that your robotics innovations could be very important to our mission."

"Thank you, sir."

"Oh, Ms. Cooper," I said as she walked through the door. "This is our robotics expert, Ms. Grason."

"Yes, we met at the party," said Ms. Cooper.

"Very good," I said. "May I get you some coffee?"

"No, thank you. Why did you call me away from my test duties?"

"We have a sample in the isolation chamber for you to examine."

"What kind of sample?"

"I think it is the powder that disappeared in your previous experiment. I think that we should get to your station and take a look. Ms. Grason, I'd like you to join us. Mr. Anderson, if you will see to things while we're gone."

I did most of the talking on the way to the examination chamber, and I could see Ms. Cooper was anxious to get another chance to examine the powder. When we got to the examination chamber, Ms. Cooper pounced on the control terminal and started the procedure.

"Where is the sample?" she asked.

"The sample was taken by one of the robots," said Ms. Grason.

"The robot has a small scoop that closes around the sample and is released and transported to the chamber."

"Where is the scoop?" Ms. Cooper asked Ms. Grason anxiously. "There is an access port to the right," said Ms. Grason. "Just punch in AP1."

She did what Ms. Grason instructed. The port opened, and there was the scoop. Ms. Cooper put her hands in the control gloves for the manipulator. "How do you open it?" she said as she turned her head and addressed Ms. Grason.

"There is a black button on the side of the scoop. Push it." She executed the move, and the container slowly opened.

"Where is the sample?" she shrieked while gazing on an empty container.

Ms. Grason and I looked at each other in disbelief since we both observed the sample being collected. My personal communicator beeped, so I pressed the response button.

"Yes."

"Sir, this is Anderson on the bridge. We need you up here, sir."

"I can't come up now, Mr. Anderson. We've got a problem here."

· "This might be a related problem, sir."

"Very well, Mr. Anderson. We'll be right up. Grason, Cooper, come with me to the bridge."

"What's wrong, Captain?" asked Ms. Grason.

"I'm not sure, but it may have something to do with our problem."

At an urgent pace, it didn't take long to get to the bridge. "Mr.

Anderson, I hope you have something important to say."

"I believe you will think this is quite important, sir." He punched up the external camera that scans the top of the ship.

"My god," I said. "There is no sign of the powder."

"No, sir. Not a trace," he replied as he looked around, looking for comments. Everyone was stunned. For what seemed like an eternity, there was silence. We just stood there, unable to express anything and almost afraid to break the silence. Finally, I asked the ladies if they could use a cup of coffee.

"Yes, with a little bit of bourbon," said Ms. Cooper. "The same for me," echoed Ms. Grason.

"I can't do that. We are on duty."

"Black then," they both said almost in unison. "Mr. Anderson?" I inquired.

"Yes, sir, also black."

"Are you sure you saw the sample being taken?" asked Ms. Cooper.

"Run a playback of the sample collection, Mr. Anderson," I requested.

He punched in the code, and the screen showed what we had seen.

"There, Ms. Cooper. See the powder covering the ship?"

"Oh my," she said.

"See the robot coming from the right?"

"Yes, but it is too far away to see the sampling."

"Just before we took the sample, we switched to the robot camera. There is where we switched. You can see clearly the robot taking the sample."

"Yes, I see."

Again, silence. I was at a loss to give orders, so I served the coffee. I could see everyone was trying to think of something to say, and Mr. Anderson broke the silence.

"The testing crew should be back in a moment with Mr. Barstow."

"Yes. Thank you, Mr. Anderson," I said and then took a sip of coffee.

Everyone else took a sip also, and we returned to the strange awkwardness. I had never been more at a loss for words, and apparently neither had my crew. It must have been a strange sight, all of us sitting there sipping coffee in silence.

I could hear the crew returning, and I stood up to greet them. Everyone was engaged in conversation while walking down the corridor. Mr. Barstow led the entourage, and it was he that I addressed first.

"Mr. Barstow, how did the test go?"

"It is as I suspected. No malfunctions detected, sir."

"Well, Mr. Barstow, we have more important business at hand. I want all of the medical staff, Officer Cooper, Officer Grason, Officer Anderson, and Barstow to remain here. Everyone else, return to your duties."

They filed out quickly, and I addressed the remaining group. "We have stumbled across an interesting phenomenon. For a brief period of time, this ship was covered with a fine yellow powder."

"For a brief period of time?" asked Mr. Barstow.

"I would answer that question, but to avoid repeating myself, I have scheduled a meeting in one hour for all crew members."

CHAPTER 12

Everyone arrived on time, and they took their seats quietly. "Ladies and gentlemen, while most of you were performing Procedure 19 tests, officers Grason, Cooper, Anderson, and I witnessed an amazing anomaly."

"Before we get started, sir," said Barstow, "may we get some coffee?"

"Certainly. Everyone, help yourselves. I'm sure you will need it."

That last comment drew a few odd looks. They all prepared their drink of choice and returned to their seats.

"As I was saying, we witnessed something unusual. Mr. Anderson, while on duty, discovered our ship was covered with a fine yellow powder. We sent out our robots to collect a sample, and when we examined the sample scoop in the isolation chamber, it was empty." I paused to take a sip of coffee, and I noticed that almost everyone else did the same.

And then Mr. Barstow spoke. "Did you take another sample?"

"No, because Mr. Anderson informed us that the substance on the outer hull had disappeared."

"Did you say disappeared?" inquired Mr. Barstow.

"That's correct. Punch it up, Mr. Anderson," I said and turned to the screen. "You can see there is no powder on the outer hull at present. Now punch up a replay from the beginning, Mr. Anderson." When the screen changed, everyone responded in amazement.

"Jump ahead to the sample taking, Mr. Anderson," I said. The robot appeared on screen, and then the screen switched to the robot's camera. Clearly the sample could be seen as it was scooped up by the robot.

"There we have it. Comments, anyone?"

"Sir, how long did the sampling process take?" asked Mr. Barstow.

"Just under twenty minutes. Actually, nineteen minutes and forty-three seconds."

"Ms. Grason," he continued with a sense of urgency. "What examination facility does the robot have?"

"Plato has a lens turret on the camera with four different microscopic powers."

Mr. Barstow stepped quick to our main console and asked Ms. Grason to join him.

"Do you have liners for the scoop to prevent sample contamination?"

"No, we just use different scoops."

"How long would it take for you to fabricate a lining for the scoop?"

"From what?" she asked. "The new polymer P-405."

"It should take about two minutes."

"Well, step over here and start the process. Punch in two twenty-five."

"Where will we get the sample?" came a voice from the back of the room.

"Is that you, Mr. Decker?" I asked. "Yes, sir."

"Very good, Mr. Decker," I said, followed by a bit of laughter from the crowd.

"There may be some of the powder floating around still," Mr. Barstow replied.

"Are there any active probes on the robot?" Mr. Barstow asked Ms. Grason.

"Yes. One electrical simulator and voltage-sensor probe. A laser scanner and a fiber-optic spectral probe."

"Are they all in place and ready for service?"

"Yes, sir."

"What is the minimum scan time for the full-probe array?"

"Three point one seconds."

"Then program the robot for full analysis and camera view."

"Are they all in place and ready for service?"

"Yes, sir."

He started the process and rapidly entered the codes.

"I want everyone except Ms. Grason to return to your station and stand by," I announced.

"If you will stand ready for some fancy ship maneuvers, Captain?" said Mr. Barstow.

"Certainly, Mr. Barstow. Whenever you are ready." I sat down and prepared for the procedures. Mr. Barstow was in his element, and it was a joy to see him in action. Ms. Grason was also impressive, and the two of them worked well together. I sat back waiting for Barstow to give the go-ahead.

"Ms. Grason," Barstow asked. "Do you have the liner completed?"

"Yes, sir."

"I need you to place a micro-impact detection strip between the lining and the scoop. I will help you with the procedure."

"The liner is ready, sir."

"Very good. Now the robot can install the detection strip. Just punch in DS-949 robot."

The robot appeared on the screen and moved over to the R-79921 panel. He placed the scoop in the opening, and the detector strip was fixed in place. The robot withdrew the scoop and turned to face the fabrication module. Ms. Grason punched up the instructions to install the lining.

"We are just about ready," said Barstow. "Prepare to maneuver the robot into the air lock." She started programming instructions.

The robot entered the air lock, and the hatch closed.

"Activate the robot's camera," ordered Barstow.

Ms. Grason punched in the code and continued with other instructions. Skillfully, she jockeyed the robot to the proper position and entered the lockdown code.

"Raise the scoop to face forward and lock it in place. Set the camera and all probes on standby. Captain, it's your tune now. I have an impact monitor scale on the top of the screen. I want you to go into a spiral course until the impact scale reads at maximum."

"For what duration?" I asked.

"Exactly twenty seconds. If we can't collect in that time, we can forget it."

"Very well, Mr. Barstow. Are we ready?"

"At your discretion, sir."

I stepped to the intercom button and made an announcement to all stations.

"To all crew members. We are about to initiate some course maneuvers that would normally be impossible. Our new IGS system, however, enables the most incredible maneuvers with little or no inertial consequences. I would advise that you stay put until the maneuver is complete. Your terminal screen will go red. When it returns to normal, you may resume your normal activities. If anyone notices any inertially abnormal events, please notify Mr. Barstow or me. If everyone is ready, we start in ten seconds."

The ship started its spiral path, and the stars began to spin. It was difficult to focus on the impact scale, but focus I did. And after about sixteen seconds, the impact meter hit maximum.

"Lock on to normal course, Mr. Anderson," I ordered. "After just sixteen seconds. How about that, Mr. Barstow?"

"Excellent, Captain. Truly excellent. And the sample is still in the scoop. Put the robot camera back on the screen, Mr. Anderson."

The screen lit up with what looked like a bright-yellow powder.

"Take the light down to half, Ms. Grason."

"There is no light on the subject, Mr. Barstow."

"The substance is generating the light?"

"Yes, sir."

"Filter it then."

"Yes, sir."

"The illumination is incredible."

CHAPTER 13

The examination continued with lens and filter changes and different sample treatments. Barstow was working at a frantic pace, programming information and manipulating sample fragments. This went on for over fourteen minutes, and suddenly…

"Gone. It's gone!" yelled Mr. Barstow. "Check the weight, Ms. Grason."

"The weight of what?"

"The weight of the robot."

She punched in the code, and the weight of the robot appeared on screen.

"Now call up the weight of the robot prior to entering the air lock."

She again entered the data, and the screen lit up with the new information. Mr. Barstow, occupied with other programming, asked Ms. Grason if the weight was the same.

"No, sir, it is not."

Before I had a chance to say anything, something impacted the ship with tremendous force. The hull-breach alarm sounded around the ship.

"Mr. Anderson, locate the breach!" I shouted.

"Deck 7, sir, back by the ion engines. Starboard side."

What now? I thought while running down the corridor. I had six decks to cover, but it was all downhill. I hadn't run that since my track-team days at the academy. I was just hitting top speed when I saw officers Yates and Dennison helping Engineer Kile Basser. I could see the severity of the injury. Mr. Basser was lying there, unconscious and bleeding profusely from the stump of a severed leg. The medical officer was doing everything she could to stop the bleeding. I picked him up and headed to the medical lab.

"How did this happen?" I asked. "I'm not sure, sir," said Ms. Yates. "Well, let's get him down to lab 3."

The two women ran ahead to prepare an examining table. I entered the lab and put Mr. Basser on the table. Ms. Dennison was getting her laser equipment ready as Ms. Yates held a bloody rag to the leg. "What is that?" I asked.

"It's Ms. Dennison's lab coat."

"Hold it tight, Ms. Yates."

Ms. Dennison returned with her equipment and started to work.

"I'll check back with you, Ms. Dennison. I must get back to the bridge."

"Yes, Captain. Could you page Ms. Kane and ask her to join me?"

"Yes, Ms. Dennison. Right away."

I stepped to the console and made the page. I quickly ran back to the bridge. I also had a sick feeling about what I would discover when back up top. The first thing I did was to pour a cup of coffee and wait for the cascade of comments. They were all so busy that they didn't even notice I had returned. Then I saw Ms. Yates just outside the doorway. She walked in and whispered in my ear, "Officer Dennison has stopped the bleeding and stabilized the patient."

"That was rather quick, wasn't it?"

Just then, Barstow saw me standing there. "Captain. We are going to be a while processing the information so if you have anything important to do, now would be the time."

"As a matter of fact, I do. Mr. Barstow, you're in charge of the bridge." Ms. Yates and I slipped away quietly. When we were down the corridor, I started questioning her.

"Ms. Yates, a man lost his leg, and you are treating it rather casually."

"Well, it doesn't seem to bother the patient. He was awake, talking and even laughing before I left sick bay."

"Well, let's get back down there. I have been back and forth so many times I feel like a yo-yo."

"You have been scurrying about a lot lately."

"Don't let it get around. I don't want my crew to think of me as scurrying around. It's just not dignified. Or captainly."

"My lips are sealed, sir."

"What happened after I left sick bay?"

"Ms. Dennison stopped the bleeding quite easily and then the patient opened his eyes and started talking. He showed no sign of pain or shock."

"Have you been on the handball court lately?"

"Actually, I was hoping to get together with you for a round of golf."

"I didn't know you played."

"Yeah, I won last year's tournament at the academy."

"Now why didn't I know that?"

"Damned if I know," she said with a smile.

I smiled back, and we continued on. When we reached sick bay, I could see that Dennison was anxious to speak to me.

"Captain, I can't explain what happened. For such a serious injury, he has no symptoms of shock. His blood pressure is barely off normal, and I didn't even give him a sedative. It is not natural."

"Is there anything you can think of that might account for this?"

"Short of being an alien? No, and I assure you that no human could be that unaffected by the loss of a leg."

"Very curious. I want you to keep me updated on any developments, and I mean anything."

CHAPTER 14

I was tired and returned to my quarters. So much had happened in that last few hours. I didn't anticipate sleep coming too easily. Maybe one of those concoctions of Officer Kane's would do the trick. I thought, Now where did I put that container?

"In the lavatory, of course," I said aloud in a manner reminiscent of one of my old school teachers.

I suddenly realized how important the minutia of everyday life was to maintaining sanity. Maybe the contemplation of easily solved problems offsets the frustration encountered with complex problems that are not so easily solved.

For some reason, I thought of pizza. I suppose because there was this little restaurant near the academy that made the best New York-style slices I had ever eaten. We had the proper ingredients, but it's just not the same. There was something about Earth pizza that was totally unique.

As I suspected, the thoughts of past pleasant trivialities served as an elixir for sleep. Beautiful visions passed through my mind, and never had I seen in a dream shapes and colors that might have been more the product of an artist's imagination. I was swept away in a virtual kaleidoscope of wonderment.

I woke refreshed and ready to go to work. I splashed water on my face, dressed quickly, and headed to the bridge. Everything was uncommonly quiet, and at first, I didn't think anything of it. I pressed my person communicator but did not receive an answer.

There was no one on the bridge, and I panicked and started running around in search of anyone. Each station was abandoned.

Where could they be? I ran by the conference room, and the door slid open.

Mr. Barstow spoke, "Captain. Everyone is waiting for you. Did you forget the meeting you scheduled?"

"Of course, Mr. Barstow. I just had a last-minute complication. Is everyone here?" I asked as I tried to gather my thoughts and not look as foolish as I felt.

"Yes, sir. We are all here."

"Well then, let's get to it," I said with something less than confidence in my voice.

"Would you like some coffee, sir?" he asked, seeming to know that I could use a pick-me-up.

Mr. Barstow walked to the coffee machine and poured a cup.

I could tell that he suspected something. Still he gave no sign to the crew that anything was wrong. He brought me the coffee and then started the meeting.

"The official results of Procedure 19 is that all circuits have tested full functioning."

"Then our ship is not falling apart?" I said, and everybody laughed.

"On the contrary, Captain," said Barstow, "the ship is performing at its peak."

He turned to the crew and started telling about the experiences on the bridge.

"While you were performing your duties, we were having an interesting adventure on the bridge. Mr. Anderson discovered a powdery substance on the outside of our ship."

He stopped for a sip of coffee and continued, "The substance disappeared within one hour. An attempt to collect more powder was successful, and intense analysis revealed the following information."

He punched up the information on screen and read it from the computer monitor.

"The powder is a unique amorphous substance that combines

with whatever it touches on a molecular level." He paused and looked up at a stunned crew. "Now I know you all have questions since this defies certain laws of physics that we have come to know, but we continue to study and evaluate the substance."

"But, sir," came a voice from the back of the room, "if it combines with everything it touches, how can you study it?"

"Is that you, Mr. Brock?" I asked. "It's you that I am always hearing from the back of the room."

"Not always, sir."

Everyone chuckled, and for a moment, the levity broke the tension that hung in the air. Mr. Barstow brought us back to reality when he spoke. "In answer to your question, Mr. Brock, while we had the power isolated, we tagged it with a tracer. This enables us to monitor the activity of the substance even after it has combined with anything else."

"Is the substance safe? Could it be a danger to us?" said Officer Kane.

"No negative effects have been observed in anything that we have studied, but we are keeping it under constant watch. We have devised some new tests that enable us to deal with the new physics the substance has brought to us."

"How do we refer to the substance?" asked Ms. Kane.

"We have given it the designation of A-1. A for amorphous, and 1 for first…"

"Sounds like something I'd pour on my steak," said Mr. Brock. Again everyone laughed, and as usual, Mr. Barstow spoke without so much as a grin.

"Well, Mr. Brock, to avoid confusion and since this is the seventh mission of the Mobius, we will name the substance A-107. If there are no further questions, this meeting is adjourned."

"Mr. Palasco," I said. "Could I see you for a minute?"

"Yes, sir, what can I do for you?"

"My personal communicator seems to be malfunctioning. Could you take a look at it?"

"I'll have a new one back to you within an hour."

"Thank you, Officer Palasco."

"Oh, Captain." said Ms. Dennison. "How is our number 1 patient?"

"In no apparent distress. Sleeping peacefully. A condition I cannot explain."

"Is there any chance of reattaching the limb?"

"No, sir. There was too much damage to the leg."

"What do you attribute his calm to?"

"I looked up Mr. Basser's records, and I found that he practiced transcendental meditation. But I can't see that would account for this calm state."

"Keep me informed about his condition, Officer Dennison," I said as I saw the perplexed look on her face. She turned and walked away, shaking her head in confusion.

CHAPTER 15

My father taught me that worrying about anything was not a good idea, but worrying about things that are going well was utter madness. So I got up early and headed to the golf course. When I arrived, Ms. Yates was waiting for me.

"You said that you would teach me," she said, smiling.

"What could I teach the current champion of last year's academy tournament?"

"Don't be so modest. I hear you are nothing short of great."

"I don't know about that."

"What do we do now?"

"We select a course. How about Pebble Beach?"

"You have Pebble Beach in the computer?"

"We have every major golf course. Would you prefer another?"

"No. Pebble Beach golf course is just fine."

I pushed the Select button, and the holographic generator placed us in position to tee off. I set my feet, took a deep breath, let it out, and swung at the ball. A 220-yard drive straight down the fairway was my reward.

"Now step forward, Ms. Yates. Take your club and tap on the floor."

A balled popped up, and she took her position. She took a deep breath, let it out, and swung at the ball. Straight down the fairway for 230 yards were the results. She played extraordinary golf and left me in the dust.

When we finished, I suggested we have coffee. She accepted, and we sat for a while, quipping.

"Do you always strive to prevail in athletic competition with your commanding officer?" I asked.

"I always play to win."

"I'll just bet you do."

"You almost won the handball game."

"Yes, I almost won."

"I think you let me win."

"I suppose you think I deliberately played less than my best on the golf course?"

"You might have."

"I assure you I did not. Why would you think I would?"

"You're a gentleman. Don't you understand? I can never win without wondering if you let me win."

"That must be frustrating."

"Not really. It's nice to know that you'd do that for me. And besides, I enjoy the game, and you are tough competition even if you deliberately lose. You keep me at my peak."

"Interesting," I said.

"So I can't win even if I win, and you can't lose even when you lose."

"How can you enjoy that?"

"It's just a game. Besides, it's kind of fun to try to figure out whether I won or you deliberately lost."

"But I wouldn't lie to you."

"Not about anything important. But like I said, you're a gentleman."

She slung her sweater over her shoulder and walked away. I turned off the holographic image and said aloud to myself, "That young woman has style."

Lost in thought, I was startled when Ensign Clark walked up behind me.

"I'm sorry, sir. Did I startle you?"

"That's all right, Ensign. What can I do for you?"

"Sir, I was asked to find you to tell you that your presence is required on the bridge."

"Is something wrong, Mr. Clark?"

"No, sir. Mr. Barstow just wants you to participate in an avoidance procedure."

"Very well. I will be there shortly."

I went to my quarters, took a quick shower, dressed, and with haste, made my way to the bridge. My presence was hardly noticed when I stepped onto the bridge. Everyone was busy at their station, engaged in serious activity.

"You requested my presence, Mr. Barstow?"

"Yes, sir. We have a situation developing. It's an asteroid grouping ahead, and it has been diverted by the gravitational pull of Neptune to a trajectory that intersects our present course."

"Can't we move around it?" I suggested.

"It's approaching at enormous speed due to the trajectory change," said Barstow.

"Where is Mr. Anderson?" I asked.

"Sick bay, sir," said Mr. Blake. "Some kind of visual problem."

"Well, Mr. Blake, it's you and me. Mr. Barstow, how long do we have?"

"About three hours," he said calmly.

"Well, Mr. Blake, get ready for a wild ride."

"What is the average size of these asteroids, Mr. Barstow?"

"Roughly the size of Mount Everest."

Mr. Blake punched up the intercom and looked up for further orders. "Attention, all stations. This is your captain speaking. We are going to execute some evasive maneuvers that might cause problems even with the IGS system. Tie everything down that you can and sit tight. We have about three hours to get everything secured including yourselves. You may view the maneuvers on your terminal screen."

I stepped behind Mr. Blake and put my hand on his shoulder. "Mr. Blake. I will give you course changes that you will enter in the flight computer. You will not engage until I say now. Understand?"

"Yes, sir," he said confidently.

"Well then, Mr. Blake, set response time to variant mode, proximity sensor override, pitch and yaw set to MRT system. Deflector shields set to VIR buffer control. Set primary ion drive system on standby. Full strut and probe retract. IGS system set at MIC. All right, Mr. Blake, standby to initiate…Engage."

On our main screen, a small object was visible on screen. "It looks like they are ahead of schedule," said Mr. Barstow.

"The origin, Mr. Barstow?"

"Not of this system, sir. The cluster has entered our system as a result of some disturbance."

"Is it likely to present a problem to our system?" I asked

"No. At the present trajectory, it will pass within 26 million miles of Uranus and then exit our system in about nine years."

"Very good, Mr. Barstow."

The movement of the cluster seemed slow and undulating. It was a stunning sight that was both menacing and hypnotic. Like a moth to the flame, we were drawn closer to the cluster.

"Are there any gravitational considerations, Mr. Barstow? And their location?"

"They are the last we will encounter."

"That's unusual."

"To say the least, Captain."

"Can you give me an estimate of the largest?"

"About forty miles in diameter. We have about one minute and twenty seconds to maneuvers, Captain."

"Do you hear that, Mr. Blake?"

"Yes, sir," he said with just a bit of nervousness in his voice. "Get ready, Mr. Blake. Alter our course by ten degrees starboard with a five-degree positive pitch change. Now."

The ship positioned itself to slide between two of the giant rocks. As we passed by, the sight was amazing. Alternate smooth and cratered surfaces spoke of a violent history from another galaxy.

"Surface scan, Mr. Barstow. Surface is extremely irregular, indicating gigantic collisions. There are some elaborate cave systems."

We were positioned to see the next surface, and it was more spectacular than the first.

"Cut our speed by 20 percent ion power, Mr. Blake. Make a ten-degree course change to port and a port yaw of seven degrees. Now."

CHAPTER 16

The task before us was becoming more difficult, and I noticed everyone was so entranced that they weren't concentrating. "Keep about it, everyone," I said with authority. "Mr. Blake, we need to make a thirty-five degree turn to port and a starboard yaw of twelve degrees. Now."

"Sir, there is a large one coming up on the starboard side."

"I see it, Mr. Blake. Change to nineteen-degree port yaw now."

"Captain," said Mr. Blake with a touch of panic in his voice.

"Standby, Mr. Blake," I said in a reassuring tone. "Keep those coordinates."

We sailed between the two closely aligned asteroids.

"Give us seventy-degree starboard roll time of 3.5 seconds," I said while looking at the mapping screen on my console. He programmed the navi-computer and waited for my order. I gave him a reassuring squeeze on his shoulder.

"Steady, Mr. Blake. Steady. Steady. Now."

Our screens filled with the dizzying sight of two huge asteroids turning just before we slipped through the narrow separation.

"Turn our ion power down to 5 percent, Mr. Blake."

Sighs of relief came from around the room, followed by applause. "Calm down, everyone," I said. "We're not out of trouble yet." Straight ahead was an enormous rock beginning to have a gravimetric pull on our ship. "Is that the largest just ahead, Mr. Barstow?"

"Yes, sir. It has an elaborate cave system."

"Any way we can explore it?"

"Do you mean a fly through?"

"Yes, Mr. Barstow, a fly through."

"No, sir. Too small."

"Then prepare a probe, Mr. Barstow. I'll get us in position for firing if you will punch in the coordinates."

He quickly prepared a probe for launch and locked down a flight trajectory.

"Ready for launch, sir."

"The coordinates are coming up on the screen."

"Very good, Mr. Barstow. This will be our final maneuver, Mr. Blake. Starboard yaw of five degrees and set ion power at 40 percent. Now. Probe launch in five, four, three, two, one. Launch. Thank you, Mr. Barstow. Now give us primary power, Mr. Blake. Now let's get out of here."

"Great flying, Captain," said Mr. Blake. "Yes, great flying, sir."

"Is that you, Mr. Langer? I didn't know you were on the bridge."

"Yes, sir, it's me. This was as good as skiing, sir."

"Well, we try to have something for everyone."

"I think this deserves a toast."

"You may be right, Mr. Langer. Then it's back to work."

The customary groans came from around the room from the more boisterous jokers in the crew. We broke out the wine, and Mr. Blake proposed a toast.

"To Captain Jenner, the best captain in the fleet." The cheers came first, then all raised their glasses and sipped the vintage.

"Thank you for your kind words, Mr. Blake. I would like to thank you for your fine navigation, and I would like to take this opportunity to toast the entire crew, the best crew in the entire fleet."

Again cheers could be heard around the ship, and we were definitely in the midst of a major celebration.

"I hate to bring this gathering to a halt," I said. "But we have a lot of work to catch up on, so everyone, back to your stations."

A few moans and everyone dispersed. Mr. Blake took a final

gulp of wine and stood to face me.

"Mr. Blake, you have earned a rest, so you are dismissed."

"Thank you, sir." He turned and left the bridge.

"Mr. Barstow, take over the bridge," I said and left.

The crew had gathered just outside and was engaged in conversation. "Ladies and gentlemen, I thank you for your accolades, but it's time we all returned to our assignments."

"I think I speak for the entire crew when I say that we need just a little more time to wind down," said Mr. Langer.

"Very well. One hour, and then I expect everyone to be back at their stations."

I could see Ms. Yates coming through the crowd. "Captain. I saw the whole thing on my terminal screen."

"You mean the flight maneuvers?"

"Of course. It was incredible."

"Just another day at the office."

"Stop that."

"Stop what?" I asked.

"Stop being so casual about this."

"Should I be formal about it?"

"You know what I mean."

"Ms. Yates, I rarely know what you mean. Now come over here and have a glass of wine."

She joined me at the table. "Why are you being this way, sir?"

"Being what way?"

"You're not taking me seriously."

"I certainly do take you seriously. I've seen you on the golf course."

"That's not what I mean. Now stop that."

"You're right, Ms. Yates. I'll stop it. Now let's enjoy the wine."

"You're raining on my parade. A parade in your honor."

"Ms. Yates. All aboard this ship, you included, are capable of doing some amazing things, things that may not appear as spectacular as the flight maneuvers but are every bit as important."

She sat there like a child that had been scolded. For a moment, she thought about it and then smiled. "You are a remarkable man, Felix Jenner."

"That's Captain Jenner to you, Ms. Yates."

She got up, walked to the doorway, turned back, and smiled again.

"That's Communications Officer Yates to you, Captain."

CHAPTER 17

I sat for a few minutes, enjoying my wine and savoring the moment. Ensign Clark entered the room and walked over to me.

"Captain. That was fantastic, sir. I watched on the monitor in communications with Ms. Yates."

"Thank you, Mr. Clark. It was quite a ride."

"The best…sir."

"Care for some wine, Mr. Clark?"

"Thank you, sir."

I reached for the bottle while the ensign grabbed a glass and stood at the other side of the table.

"Have a seat, Mr. Clark."

"Yes, sir."

"You were impressed with that, were you?"

"Oh yes, sir. I've never seen anything like it."

"Well, neither have I."

"It was exhilarating."

"Where are you from Mr. Clark?"

"Connecticut, sir. New Haven."

"Did you always want to serve aboard a starship?"

"Oh yes, sir. Always."

"How did your parents feel about that?"

"They didn't like it too much at first, but they came around."

"I hear you are quite a musician. You've played at Carnegie Hall."

"When I was seventeen, sir."

"Did you want to pursue a musical career?"

"No, sir. I didn't."

"I bet your mother wanted you to."

"Yes, sir. She did."

"You were looking for adventure though."

"Yes, sir. I've always wanted this."

"Well, you certainly got it, Mr. Clark."

"Captain. When do you have a moment?"

I looked up to see that Mr. Barstow had entered the room. "Mr. Barstow, I'll be with you in just a moment." I turned to

Mr. Clark. "I enjoyed our talk, Mr. Clark. But duty calls, so I must go."

I joined Mr. Barstow as we headed toward the bridge. "Captain, our probe has found something interesting."

"And what is that?"

"Let's go to the console."

We stepped to the control panel, and I could tell that Barstow had found something important. After entering the playback code, he turned to me and started his presentation.

"Our probe has discovered something significant at the end of the cave in the asteroid."

"And what would that be, Mr. Barstow?"

"A mechanism."

"A what?"

"A device."

"Evidence of intelligent life?"

"You could say that, sir. The mechanism is the remains of Voyager 2."

"Voyager 2? That must be nearly a hundred years ago."

"One hundred years, ten months, four days. It was launched by NASA, the academy's old designation, on August 20, 1977. It was one of the most successful exploration satellite probes ever launched. It was expected to leave our system to explore beyond the outer rim."

"Quite a history lesson," I said.

"Our probe got quite a good look, and our computer vessel template bank made a positive identification."

"What about those asteroids? You said there was something strange about them."

"There were many strange things about them, and I would like to schedule a meeting with you and Mineralogist Cooper and myself."

"Schedule it for the beginning of her next shift, which is in about six hours. That will give me about five hours of sleep."

Despite the wine and good feelings, I was concerned about the strange rocks. Again sleep did not come easy, but it did come, or at least until I felt the hand of Mr. Clark on my shoulder. He asked me to come with him to engineering. I asked him what was wrong, but he said nothing. I told him this had better be important.

When we got to engineering, I looked around, and no one could be found. I turned to Mr. Clark, and he, too, disappeared. Our main energy converter was leaking a strange purple liquid. I punched up the station communicator and told Mr. Matson to get to engineering on the double.

More of the liquid was pouring out all over the floor, and systems were going critical all around. All I could do was watch as the slimy substance covered the floor. I hopped up on a chair to avoid contact, but it was dissolving under me. I was alone and in a hopeless situation, and alarms were going off all over the ship.

I reached over and shut off my wake-up alarm. Why was I having such nightmares? They were so vivid, and they had to stop. My scheduled meeting was at hand, and I had to pull myself together. Cold water on my face and deep breathing had never felt so good. I dressed quickly and headed to the conference room to meet with Mr. Barstow and Ms. Cooper.

I passed no one on the way, and an uneasy feeling came over

me. This is not another nightmare, I hope. I was relieved to see Barstow just ahead, and I quickened my pace.

"Good morning, Mr. Barstow," I said with unexpected exuberance.

"You seem to be in a good mood, Captain."

"I suppose I am."

"Can I get you some coffee, sir? Ms. Cooper will be here presently."

"Yes, coffee will be fine."

Ms. Cooper appeared at the door and seemed embarrassed that she wasn't the first one there. I'm sure her nervousness was due to the fact that she was a new crew member.

"Come in, Ms. Cooper," I said. "Would you care for some coffee?"

"Oh yes, sir. Thank you."

We all sat for a moment, sipped our coffee, and settled in.

CHAPTER 18

I nodded to Barstow, and he stood up to start the meeting.

"Ms. Cooper, Captain. Our encounter with the asteroid cluster has created quite a stir."

"Yes, sir. I recorded the pass from my terminal," she said. "And what did you conclude?" asked Barstow.

"A number of anomalies," she said, a little more relaxed after taking a sip of coffee.

"Precisely," said Barstow, pleased with her observations. "What?" I asked. "Did you find anything unusual about the asteroids?"

"Well, first," she said, then paused to consider, "the size differential is unusually large and out of order."

"Out of order?" I said.

"Yes, sir. Clusters usually travel with the larger pieces up front and the smaller ones follow, or at least a random order. This cluster had smaller ones up front and without exception, increased in size."

"What do you make of that, Ms. Cooper?"

"I cannot account for this."

"What else, Ms. Cooper?"

"Surface features."

"If I could ask you, Mr. Barstow, to run a playback of Ms. Cooper's recording."

Barstow stepped to the console and entered the playback code. He turned to the screen as Ms. Cooper spoke.

"We are coming up to the largest asteroids," she said, looking up at the screen.

"Stop. You can see the smooth surface in the upper right. Moving down to the left, we see some unusual striations, and below that, there is a pitted area around the cave. These features are not easily explained." She took a sip of coffee, and I could see that she was really getting into it. "As we proceeded through the cluster, the size of the individual asteroids was increasing. Normal explosions produce debris in a more random size arrangement. Most clusters traveling through space usually have the large pieces up front with a lot of smaller pieces following. To have the pieces move by in ever- increasing size could only mean…"

"Could only mean what, Ms. Cooper?" I asked. "Manipulation."

Mr. Barstow looked impressed with her dissertation, and he started firing questions.

"Ms. Cooper. What about the spread factor?"

"As you know, Mr. Barstow, the spread factor has no effect on small objects that are pulled along by large objects. The spread factor only applies to small objects in front of large. They will either veer off and out of the gravimetric control of the larger mass or fall back and impact with the surface of larger object. Our cluster was not the product of an explosion."

"What would be the purpose of manipulating the cluster to this formation?"

"A trap."

"How would that work?"

"Only an extremely powerful and sophisticated starship with a very skilled crew and an IGS system could maneuver through. Anything less wouldn't have a chance. This cluster has not been in existence for very long, or the spread factor would have eliminated the smaller fragments in the front. Large chunks don't follow small chunks. They pull smaller objects back to them."

"Mr. Barstow, have you briefed Ms. Cooper on our discovery?"

"What discovery?" she said, looking up.

"We found, from the probe we sent out, positive identification of Voyager 2 in one of the caves of the largest asteroid."

"Like it had been trapped," I added.

She seemed startled at the information, and she took another sip of her coffee. I gave her a moment to collect herself, and then I congratulated her on her excellent presentation. She seemed lost in thought as she left.

"Mr. Barstow, what is next?"

"We have a more important problem at the moment, namely, a substance that can't be studied because it disappears. We are trying to develop a method of containment, but under these conditions, progress is slow."

"Keep me informed, but right now, I have some business to take care of in sick bay."

Things were considerably more active than they were the last time I was there. Ms. Dennison and Ms. Kane were both busy with patience. I sat patiently, waiting for of them to finish, but it did not happen soon. This gave me the opportunity to observe the bedside manner of my medical staff.

Ms. Kane was very nurturing and sympathetic in her treatment of everyone. In contrast, Ms. Dennison was more businesslike and matter-of-fact in her demeanor. I was surprised to see so many in need of medical attention. Mr. Palasco was a patient, and as he left, I stopped him for a minute to inquire.

"What seems to be the problem, Mr. Palasco?"

"My eyes, sir. Everything appears blurred. Oh, by the way, sir, I have a new communicator for you. I've got it right here."

He fumbled around for a few seconds and then found the pocket he put it in. Then he dropped it and had trouble finding it, so I reached down and picked up. "Mr. Palasco, you have dilated pupils. Get yourself to your quarters and relax."

"Yes, sir," he said as he stumbled to the corridor. Ensign Clark was approaching, and I asked him to help Mr. Palasco to his quarters. When I looked back, Ms. Kane was ready for me, so I took a deep breath and approached with caution. "Ms. Kane, I just came down to check on how things were going."

"We seem to have an outbreak of some visual problem. Four crew members have had it."

"Do you know what is the cause?"

"No, but a concoction I put together gives some relief."

"Do you have anything for nightmares?"

"So you did come here for treatment."

"I've been having nightmares."

"I have something that will suppresses dreaming, but you'll wake up with a hangover."

"That's fine. I've got to get rid of these disturbing dreams."
She handed me a small container of pills and then made an entry in her computer terminal.

CHAPTER 19

I was looking forward to a good night's sleep when I took one of the pills and laid down to rest. It was not to be though. My dreams were more active than ever. Gross images that were unsettling raced through my mind: images of the horrors of war and injuries of the innocent victims, children dismembered and crawling on the ground.

I woke with a scream and ran to the lavatory to vomit. I had never seen anything so revolting, and I wondered from what primordial sludge these images were dredged. I did not want to go back to sleep for fear of more gross images. I tried to occupy myself with computer filing, but I just could not clear my mind. This was not the time for gut-wrenching revelations. The most important decisions of my career were just ahead of me, and I did not need the distractions. My first duty visit was to sick bay to inform Ms. Kane of my nightmare. She was busy at her computer terminal when I entered, so I waited patiently. She soon noticed me and walked over.

"I'm surprised to see you here, Captain."

"I had to drop by to tell you that the pills you gave me didn't work."

"That is certainly unusual. Those pills are the best thing I have."

"I had the worst nightmare I've ever had."

"Hop up on the table."

"Now I didn't come in for an examination."

"Well, that's what you're going to get."

"You're not going to ask me to get undressed, are you?"

"Not unless you want to," she said with a smile. "What are you going to do?"

"You are such a baby. I'm just going to do a full-body scan. Now be still." She pulled a sensing unit over to the table.

"Will it hurt?" I asked.

"Unfortunately, no." She turned on the machine. It made typical medical machine noises, and a light moved back and forth over me. After a minute of this, the machine was turned off.

"There now. That didn't hurt a bit, did it?"

"Maybe a little bit," I insisted.

She turned to the monitor for the readout, and after seeing the results, she turned to me and said, "You're in perfect condition, Captain."

"Nobody is perfect."

"You're right, Captain, as usual."

"What can you do about these nightmares?"

"I suggest you stay awake."

"As always, Ms. Kane, you have the right answers."

"You know it, Captain."

"There are several of your patients with visual problems. Is it contagious? Is it something I should be aware of?"

"No, Captain. I've come up with an eye-drop mixture that gives temporary relief, and I am studying the problem. I don't believe it is serious. It may be a reaction to the IGS programming."

"You don't like the IGS system, do you, Ms. Kane?"

"It is experimental and untried in long-term situations."

"Is there anything that leads you to believe that the system is harmful?"

"It's just a hunch."

"How is Mr. Basser?"

"Remarkably well. And it worries the hell out of me."

"It worries you that a patient is doing well?"

"It worries me when a patient doesn't react normally to traumatic injury."

"Maybe there is more to transcendental meditation than you

thought."

"I'll have a look into it."

"Well, Ms. Kane, I must get back to duties. Keep me informed of Mr. Basser's progress."

"Of course, sir."

I left sick bay feeling sick that there was no pill remedy better than what I took. A glass of warm milk—oh no, not that. Some scotch before I go to sleep. Yes, that sounded much better. In fact, I might have a sip before I went on duty, but since I did not want to set a bad example with whisky on my breath, I had vodka instead.

The bridge was alive with activity. "Mr. Barstow, what's going on?"

"Our magnetometers have detected a magnetic flux coming toward us."

"What's a magnet flux?"

"An intense fluctuation of magnetic energy moving through space."

"I've never heard of such a thing."

"Neither have I."

"Isn't our hull magnetically sealed?" I asked. "Not against this intensity."

"Nothing on board is affected by magnetic energy."

"You're wrong, sir. Our bodies contain iron, and when this flux passes, it will produce internal bleeding in everyone on board. If you will take a position at console 3, Captain."

I sat down quickly and waited for instructions. All but life-support power was being routed to our magnetic shields.

"Run primary power to buffer nine and then to shields."

"Done, Mr. Barstow."

"Now set the buffers to inversion mode. This will minimize the impact of the flux."

"Locked in, Mr. Barstow."

"We don't have enough power!" said Barstow with a degree of panic in his voice that I had never heard before. "Auxiliary power online."

"Our life support is on auxiliary power!" I yelled. "We need all the power we can get, Captain."

"We will overload, Mr. Barstow."

"We must take a chance. We only need to maintain it for 2.3 seconds. Program the transfer now, Captain."

I programmed the transfer and waited.

"It's approaching fast. It will pass us in twenty-two seconds. Get ready, Captain."

I stood ready to enter the transfer, and Barstow started the countdown. "On my mark, in five, four, three, two, one. Now."

I entered the transfer, and immediately our power plummeted. Lights all over the ship flickered and dimmed. A tremendous vibration passed over the ship as we braced for impact. In a few seconds, the flux had passed, and we all looked around to see if everything was all right. I saw Barstow with a concerned look on his face.   "Well, Mr. Barstow, you were right."

"No, Captain, I was not."

"What do you mean? Everyone seems to be okay."

"Our power was drained before the flux passed. We should all be dead."

"But we aren't, Mr. Barstow."

"No, we are not. And I don't know why."

Over the intercom came the voice of Chief Engineer Matson. "Captain, we have a complete power drain, and all that is online is a subsystem that is maintaining life support at 60 percent."

"We can't have that, Mr. Matson," I said. "Is there any way we can improve on that?"

"We are lucky to have that, sir. Regeneration is going to take some time."

"How long can we operate at 60 percent?"

"We should be having problems now, Captain."

Everyone was shocked and started voicing their concern. I took charge by making an announcement over the ship's intercom.

"Everyone, into the conference room."

CHAPTER 20

The entire crew filed into the conference room, and I realized that I had to get everything organized. "Find a seat, everyone," I said. "Now listen. "We're in trouble, and we need everyone at their best. The dim lights are not for atmosphere. We are just about out of power, and I was just told by our chief engineer that it would be a while before we can get back to normal. Mr. Barstow will explain what happened and answer any questions."

They settled down and listened closely for they knew their lives depend on it.

"We detected a magnet flux thirty-three minutes before impact."

"What is a magnetic flux?" asked Mr. Anderson.

"It's like an electromagnetic pulse that follows a nuclear detonation except these pulses are continuous."

"I've never heard of such a thing," said Mr. Decker.

"No one has," said Barstow. "Electromagnetic pulses are extremely intense. Electromagnetic flux is even more intense."

"Don't we have magnetic shielding?" asked Mr. Brock.

"Our shielding is not strong enough to protect us against this intensity," said Barstow.

"We do not have magnetically sensitive equipment. We use optical cables," said Mr. Brock.

"No," said Barstow. "But we have iron in our blood, and we should all be bleeding internally."

"But the extra power to the shields," said Mr. Brock.

"Our power drained quickly when we transferred, and the shields were inactive during the time the flux passed."

"Why did that happen? We use power buffers," said Mr. Anderson.

"I can't say just now, and we won't know until we recover power," said Barstow.

"How long will that take?" asked Mr. Brock. "Again, I can't say until we have recovered power."

For a few moments, there was silence. I looked around at a stunned crew, and I realized I had to bring something to this. "Your attention, everyone. I want everybody to report to sick bay immediately. And then you are to report back here for further instructions. Now go."

Mr. Barstow stood alongside me as they filed out. I wasn't sure what to do, but I felt I should look in-charge.

"Mr. Barstow, where do we go from here?"

"I'm afraid I have never been in a situation so perplexing," he said with an unusual quality to his voice.

"Well, let's get ready for their return. Are there any machines working?"

"Just the coffee machine."

"That's good. We all will need that. If you will hunt up some paper and pencils. We are going to be roughing it. Has anyone on board done secretarial work?"

"I believe Mr. Clark once had a secretarial position."

"Thank you. Before they get back, I want to know your assessment of our predicament."

"This is the first time I have ever said this. Your guess is as good as mine. I'm accustomed to things operating to a set of physical rules. When they do not, there is no way to apply logic."

"Mr. Barstow, in the midst of the unknown, logic is still our only weapon. You're the best computer we have on board and now, the only one. We are all depending on you, and you can depend on us for anything you need."

"Thank you, sir, for your confidence," he said.

"It's because of your past performance, Mr. Barstow. You've done some amazing things, and I have no reason to believe that your best work isn't ahead of you. If there is anything I can do to help."

Mr. Brock was the first to return, and he was anxious to get to the coffee. I joined him at the conference table. "How are things going down in sick bay?" I asked, hoping to hear some good news for a change.

"They are using the portable hand scanners to save the ship's power."

"How would you describe the mood down there?"

"Very positive, sir. They seem to be on top of the situation."

"Mr. Brock, I want you to take a seat at the navigator console, and if any power goes on, make sure the navi-computer is online first.

"Yes, sir," he said and stepped to the console.

More of the crew started to file in, and they were all in need of a caffeine break. After getting their coffee, they settled down and took their seats. I was pleased to hear that my crew remained in good spirits and they were ready to work on our problems.

"Attention, everyone," I announced. "We have much to cover, and I am sure you have many questions. So let's see a show of hands. Mr. Decker, you have a question?"

"Yes, sir. How did we lose power?"

"Mr. Decker. Our science officer detected a strange magnetic phenomenon that posed a threat to all of us, and he acted to give more power to the magnetic shields so they would counteract the effects of the menacing force. We ran the power through our buffer system to avoid overload. But it failed, and it drained a lot of power through the shields."

"But it worked obviously."

"Actually, Mr. Decker, our power drained away before the force passed through us."

"Then the force did not pose a threat after all?" asked Mr. Blake. "Apparently not," I replied.

"Then we dumped our primary power source for nothing?"

"That has yet to be determined."

"But, sir, we all survived," said Mr. Langer.

Mr. Clark came into the room and stood for a moment and then spoke, "Captain. You were looking for me?"

"Yes, Mr. Clark. I'm told you have secretarial skills."

"Yes. What can I do?"

"Since our recorders are out, we need a way of keeping records of activities aboard the ship. There is a pad and pen on the desk. If you could keep the minutes of the meeting?"

"Certainly, sir." He sat down and started writing at the top of the page. "I'm ready, sir."

"Very good, Mr. Clark." I turned my attention to all crew members. "We are all going to take on some extra duties until we can restore power. If there are any good cooks out there, it's time to step up. We've got to prepare our own food."

Several people raised their hands, and I directed them to collaborate on meal planning and report to the galley. The medical staff would continue to function as normally as possible. Mr. Greyson was given the task of monitoring the air quality. Ms. Grason was with her robots, and Ms. Yates was to track down any mechanical damage that had occurred during our recent mishaps. Mr. Matson was, of course, busy in engineering with Mr. Decker, and Mr. Langer was making the necessary repairs on the main power unit. I ordered my science officer to serve as first officer and spend more time on the bridge with me.

CHAPTER 21

As usual, Officer Dennison was busy, so I took a seat and waited. In the midst of writing on her clipboard, she spoke without looking up. "I see you, Captain."

"And I see you, Officer Dennison."

"What can I do for you, Captain?"

"I need to ask you some questions."

She made a few more notations on her clipboard, walked over, and sat down. She seemed tired, but I needed answers to some important questions.

"Okay, Captain, fire away."

"I don't mean to question the decisions of my science officer, but if Mr. Barstow was correct in his evaluation of the magnetic flux force, why are we not dead?"

"Captain, I don't deal with theoretic physics. But what he said makes sense, and I'm sure he could sit down with you and prove it to your satisfaction."

"I'd just like to be sure of it."

"It's my job to take care of the wounded, but if there is anyone that I would trust with quantum physics, it's Science Officer Barstow."

"I didn't know you had so much respect for him."

"I'm just going by what you told me about him."

"Oh, great."

"Look, Captain, I can't reassure you. Mr. Barstow has the best reputation in the fleet. What else need be said?"

"You're right," I said with a measure of relief. "Now is there anything else, Captain?"

"Yes, I am concerned about the fact that our life support is operating at just 60 percent."

"I must say, I was concerned about that myself. We should be having some reaction now, but I have seen no indication of adverse effects. So I can't worry about it."

"Oh. How is our miracle patient?"

"Better than all of us."

I left sick bay feeling better this time. She was right. I should have confidence in my science officer. He was assigned to the Mobius because it was the most advanced research ship ever built. We had served together on the last five missions of the Mobius. His scientific mind had, on more than one occasion, saved us from disaster.

I was about to go off duty, and although I was exhausted, I didn't look forward to sleep. But inevitably sleep came—sleep of a more palatable nature. Again dreams unlike anything I had ever experienced. It was therapeutic, like a Jacuzzi and full-body massage in one. The quality of sleep was nothing short of remarkable.

I woke refreshed, but the lights in the corridor were still dim and served a constant reminder of our condition. We were adrift in space but for some reason, on course. So my presence on the bridge was more to boost morale. I asked Mr. Anderson where our science officer was, and he said he's in engineering and would be back any second. And as if by magic, he walked in.

"Mr. Barstow," I said. "I want to see you in the conference room." He said nothing but followed me as if he expected a chewing out. I felt it was up to me to let him know that I had not lost confidence in his ability. He went to the coffee machine and returned with two cups.

"What did you want to speak to me about, Captain?"

"Just a few questions about the magnetic flux. First, could this be a natural phenomenon?"

"No, sir."

"Would you like to explain?"

He looked up with a surprised expression on his face. I nodded my approval, and he seemed relieved to explain his theory. He reached for his coffee, took a sip, and stood up.

"As you know, all nuclear explosions produce an electromagnetic pulse. A magnetic flux is like a series of electromagnetic pulses, as if a multitude of nuclear explosions happened one after another and all from a single point.

"What intensity are we talking about?"

"At least one hundred thousand times more intense than any electromagnetic pulse we know of at 350 pulses per second."

"Is there any other evidence that this is not just a natural phenomenon that we have never encountered?"

"Yes, sir. All natural activity is expressed in exponential relationships. Mathematical formulas are designed to compare trajectories of force whether they are counter to or reinforcement types. The magnetic flux was linear. The only known source of linear signals is some form of electronic instrumentation."

"Would you say that this was an encounter with an alien life-form?"

"Yes, sir."

"Could this be a form of communication?"

"No, sir. This signal was meant to kill whatever it passed through."

"Why are we alive then, Mr. Barstow?"

"I cannot answer that. We are encountering many new things out here, and there is no way to be sure of anything."

"You have given me a lot to think about, Mr. Barstow. Right now though, our number one priority is to get power restored, so if you would help Mr. Matson in engineering, I think we can get back to normal sooner."

"Yes, sir," he said with the sound of renewed confidence in his voice.

I had accomplished what I had hoped to. If we were to get out of the trouble we found ourselves in, we would need a science officer at the top of his form.

CHAPTER 22

Our first meal prepared by our crew looked appetizing. Since we could not spare the power for a cooked meal, we were forced to have a salad. The table was set nicely, and everything looked great. Even our dimmed lights provided us with an unexpected romantic atmosphere. A flower arrangement as a centerpiece and candles added the final touch.

Other than Mr. Greyson, who obviously supplied the floral arrangement, I was not sure who was responsible for the lovely meal. I would not be held in suspense for long because in walked Lorin and Lanie Rodgers with wine and glasses, followed by Jan Cooper with a pepper grinder. I was curious to know the ingredients as I tried to get a better look.

"It's a Waldorf salad," said Ms. Dennison.

"I love Waldorf salad," I said enthusiastically. "That's why they made it, Captain"

"And how did they know, Ms. Dennison?"

"They checked your service records."

"What?"

"Just kidding, Captain."

"I've never known medical officers to have a sense of humor. In fact, I found most of them to be quite stuffy."

"Is that a fact?"

"Yes. And by the way, medical officers aren't allowed to sit next to the captain."

"What?"

"Just kidding, Ms. Dennison."

"I've never known a captain to have a sense of humor," she said, rolling her eyes.

"We're all a little stuffy."

She decided to eat rather than banter, and I was glad she did. The meal was excellent, and it served as a form for social interaction that was less formal than staff meetings. I suggested that we make this a regular activity, and everyone applauded. Ms. Dennison, not one to ignore an opportunity to flex her wit, added one final point.

"It's amazing how good everyone looks in candlelight."

The comment met with a variety of moans, groans, and even a chuckle from those less sensitive about their appearance.

"As usual, Ms. Dennison," I interjected, "you continue your crusade to win new friends." My sarcasm was accepted in good spirits, and I considered the dinner a rousing success.

I stayed to clean up, maybe pick up a recipe or two, and see how our chefs felt about their new duties. Everyone seemed to enjoy the interaction, and the laughter was infectious. I was glad to see that despite our problems, morale was high.

"Ms. Cooper, you wield a mean pepper grinder."

"Thank you, Captain. Do you think I missed my calling?"

"I'll put in a good word for you at the academy restaurant."

"It's always good to have something to fall back on."

"For the time being, we need your services as a mineralogist."

"Well, I'm glad I still have a job."

"Speaking of that, I was very impressed with your report."

"Really, Captain?"

"And so was Mr. Barstow."

"Is that a fact?"

"You've convinced me, Ms. Cooper. There is something I need to know."

"What's that, Captain?"

"Who made the salad?"

"That would be the Rodgers twins."

"Our chemists, of course. Why didn't I think of that?"

"Dammed if I know?"

I excused myself and walked over to the Rodgers twins as they were washing dishes.

"I hear that you two are responsible for our delicious salad."

"Yes, sir," said Lorin, or I thought it was Lorin. "Mr. Greyson helped us with the ingredients."

"Well, ladies, you are to be commended. I look forward to our next culinary gathering."

I noticed Ms. Kane was ready to leave, so I stopped her in the corridor. "I wanted to speak to you, Ms. Kane. May I walk you to your quarters?"

"Certainly, Captain. What can I do for you?"

"I just wanted to know how things were going in sick bay."

"Anything specific?"

"Well, for one thing, how are those visual maladies that have cropped up?"

"I haven't isolated the problem, but the eye drops are working for the time being."

"Anything else?"

"Nothing is going on now. I can't work without my equipment."

"Oh yeah. That's right."

"How long until we get powered up?"

"I can't say for sure, Ms. Kane. Mr. Barstow is down in engineering with Mr. Matson working on the problem as we speak."

"I'm using the time to catch up on my correspondence."

"Very resourceful. Give my best to Aunt Mary."

"Captain, I thought you said that commanders were very stuffy."

"Now when did I say that?"

"During dinner."

"Really?"

"You're putting me on, Captain."

"I wouldn't do that."

"Sure you would. You're just mean enough."

"You think I'm mean?" I said as if I was hurt by the cutting remark.

"Yes, I do."

"I'm just trying to keep up with my crew."

"You're doing more than keeping up. Well, I'm home."

"Yes, Ms. Kane. I'd come in for a cup of coffee, but I have such a long drive ahead of me."

"I understand. Any time you want to go back to that charming little restaurant, just give me a call."

"I don't have your number."

"I'm in the book."

CHAPTER 23

I had never imagined that life aboard a starship could be called dull. However, drifting helplessly with no power or instrumentation was anything but interesting. Food preparation, laundry, and other mundane activities gave us some sense of accomplishment until we could return to our normal duties.

To relieve some of my boredom, I decided to visit engineering and see for myself how repairs were going. Things did not look so bad. However, the smell of burned circuitry was heavy in the air. Mr. Matson and Mr. Barstow had disassembled a large panel from the main console and were examining it closely. They didn't notice my presence, so I just looked around on my own. I could see the damage was quite extensive.

Continued surveillance did nothing to bolster my spirits. Scorched panels everywhere gave evidence of a tremendous overload, and the repair would take longer than I had anticipated. Mr. Barstow caught sight of me and came over to give me an update.

"It's not as bad as it looks," he said apologetically. "It couldn't be, Mr. Barstow," I said sadly.

"Once we have made repairs, this mess will clean up easily."

"I certainly hope so. How is it going?"

"We must be finished in six days."

"If we can't regain control of our ship by then, we will pass by our destination. Do you think we can make it?"

"I think we'll be ready in five. I only hope we don't hit anything before we regain control."

"What do you mean?" I said with concern.

"Since we don't have power, our IGS system is not working. Our speed did not decrease, so if we hit anything, we'll be tossed about rather violently."

"Should I warn everybody?"

"To do what?"

"I don't know, Mr. Barstow. If nothing can be done, why tell me?"

"Obviously a mistake."

"No, Mr. Barstow, I'm glad you told me. I've not had enough to worry about lately."

"Yes, sir."

It was terrible to wish you didn't know something you were glad to know. I made my way through the rubble and back to my room where Ms. Yates was just about to press my doorbell.

"Ms. Yates, can I do something for you?"

"Captain, the atmosphere monitor has dropped to 55 percent."

"Are you sure?"

"Yes, sir."

"Come with me to sick bay, Ms. Yates."

We proceeded quickly, and all the while, I was thinking to myself, What else can go wrong? I realized that I didn't want to hear the answer to that question. Ms. Kane was sipping coffee when we entered, and I wasted no time expressing my concern about our atmosphere.

"Ms. Kane. I've been informed that another 5-percent drop in oxygen has been recorded."

"What do you think I can do about that?" she said, frustrated. "I just thought you could advise us."

"I didn't think we could function at 60 percent. That's like camp 3 on Everest, and now you tell me that it has changed to fifty-five. I've had no respiratory problems reported, so I can't worry about it."

"I just thought you might have some ideas."

"It is difficult to form ideas when nothing is responding as it is supposed to. Now if you have no further questions, I have some eye drops to prepare."

"How is the remedy working?" I asked.

"The symptoms are getting worse, and it is difficult to work without a computer."

"We're trying to correct that as we speak. We'll let you get back to work now."

Ms. Yates and I walked to the conference room. I poured two cups of coffee, and we sat down for a few minutes.

"Are you okay, Captain?"

"No, none of us are. As long as we remained mired in these problems."

"We are all doing what we can."

"I feel so helpless."

"There is nothing you can do, Captain."

"Maybe there is," I said with sudden inspiration. "Ms. Yates, let's get down to engineering."

We almost ran through the corridor, my mind racing with ideas.

Mr. Barstow was surprised to see us back.

"Ms. Yates," I said. "Grab a few towels from the utility cabinet." She stepped over to the cabinet and returned with a handful of towels. "Clean panel number 4," I said. "And throw me one of those towels."

For the next few minutes, we cleaned and replaced panels. Ms. Grason appeared at the door. "Need any help?" she asked. "Sure," I said. "Come on in."

"I brought Archimedes with me," she said cheerfully. "What can he do here?" I asked.

"With the new devices I've designed, quite a lot."

"Well then, we are glad to have him."

The robot came in and immediately started beeping. "What is that, Ms. Grason?"

"He's speaking robot, Captain."

"What is he saying?"

"He is asking for orders, Captain."

"All right. Install panel 4 in place," I said and watched as the robot moved to panel 4, picked it up, put it in place, and adjusted the retaining clamps. He backed away and made a strange sound, like an audio sweep.

"What was that?" I asked.

"He just told you he was finished," she said. "Well, isn't that great?"

"Now say his name, Captain."

"Archimedes, install panel 9," I said.

He completed the task in record time and once again performed the audio sweep.

"That's very helpful, Ms. Grason."

CHAPTER 24

For the next few hours, we all worked efficiently, and when we had everything cleaned up, we started helping with repairs. The robot could do spot welding, laser fusing, and component testing. Mr. Barstow seemed pleased with the help, and an idle crew was more than happy to lend a hand.

All this activity would speed up recovery time, and the boost in morale was certainly welcome. Archimedes was efficient and quite amusing.

"Ms. Grason," I said. "Your additions to the robot are quite unique, and I think I want to adopt him."

"Thank you, Captain. I think he likes you too."

I left everyone busy at work to alert the bridge to testing procedures. Mr. Anderson was on duty, and he was replacing a panel on the navi-computer.

"Mr. Anderson," I said. "Get ready for tests in about fourteen minutes."

He secured the last panel clip and returned to his seat. He engaged the test module and waited for the order to test display. The buffer and isolation stages were the first, and one by one, each section tested good. Next was our interface matrix, and again everything was fine. Our life-support system was operating at 50 percent, and all the warning lights were glowing.

I pressed my personal communicator and called, "Mr. Barstow."

"Yes, sir?"

"Could this be a false reading?" I asked.

"No, sir. All of the portable units read the same."

"How do you account for this, Mr. Barstow? Ms. Kane tells me that we should not be able to function at this level."

"She is right, sir. And I can't explain it just now, but I am working on a theory."

Our primary engine system was next and then our ion engine array. After that came the IGS system, and the warning light came on again.

"What is the problem, Mr. Anderson?"

"Nothing, sir. We don't have enough power to operate the IGS system."

"We do have enough power to accommodate the ship's intercom," announced Mr. Barstow over the ship's speakers. "We need to test the external relay and light system, so to prevent overload, we're going to shut down all internal lights. We will do this in one minute, so sit down for three minutes and don't move around."

We all followed instructions and sat still for the three-minute test period. The darkness was eerie and unsettling. It seemed much longer than three minutes. Then the lights came on, and I breathed a sigh of relief. A series of subsystems, medical and lab instrumentation, and food processing were next on the list. Camera systems and environmental thermal control were to follow. Our external and internal sensor system could not be tested until our mainframe was fully back online. The main engine power was also not accessible until the mainframe was fully restored.

Mr. Barstow announced that we could expect full power in forty-eight hours. Life was getting back to normal. Our lights were brighter, the oxygen level had returned to normal, and everyone was anxious to return to duty.

I visited engineering to congratulate Mr. Barstow and Mr. Matson on their excellent work and found the premises spotless. He told me that some of the ladies came down and cleaned up the place. Looking around, I could definitely see a feminine touch. On my way back to quarters, I stopped by sick bay. Ms. Kane was sitting at her station, looking up at the ceiling.

"Daydreaming?" I asked.

She turned to face me and asked, "Is it possible to daydream in space?"

"It's possible to daydream at night," I replied.

"Then I guess I was daydreaming. How can I serve you, Captain?"

"When our computers come back online, I want you to check something. During our power-down time, our atmosphere monitor registered as low as 55 percent normal oxygen. I want you to run a check of all aspects of air evaluation and what the exact tolerance parameters are."

"Just when will that be, Captain?"

"In about forty-five hours. And barring medical emergency, this is a priority. Let me know the moment you get the results."

"It will be done, sir."

"You obviously want to get back to work."

"I certainly am."

"I've been meaning to talk to you about your first deep-space mission, but with all that has transpired…"

"I understand, Captain. We've all been through the mill."

"If you'd rather do this some other time?"

"No. I didn't mean to put you off. I've been doing some tests on the IGS system."

"That's right, you're not a big fan of the IGS system."

"No, sir, I'm not. But there may be a connection between the IGS and our ability to function under somewhat less-than-ideal atmospheric conditions."

"And what brought you to that conclusion?"

"The only thing that we have all been exposed to for months

now is the effects of the IGS system. Although I don't believe it is responsible for the immunity to adverse conditions, I can't ignore the possibility."

"That's very interesting, Ms. Kane. I look forward to reviewing the final results of your research. And let me ask: are you satisfied with the facilities?"

"Oh yes, Captain. The facilities are even better than on the Stockton. If I seem depressed, it's just the problems we've been having and the extra duty that I have in sick bay."

"Ms. Kane. You have been taking care of all of us, and I think you should get some needed sleep and relax for the next forty-five hours. You need a break, and now is the perfect time to take it. And that's an order, Officer Kane."

"Yes, sir."

CHAPTER 25

My entire crew was performing brilliantly, and I wanted them to continue for just a little while longer. Mr. Barstow approached me on the bridge. "Captain, we have our sensors online, and they have detected a field of small asteroids just ahead. And we can't avoid them. We do not have the IGS unit back yet, so we must tie ourselves down."

"How long do we have?"

"Just twelve minutes."

I slammed my hand down on the ship's intercom control. "Attention, all crew. Get to your stations immediately and belt up. We are about to enter a field of small asteroids and should receive a rather heavy pelting." To make sure that everyone was alerted, I activated the emergency alarm. The crew on the bridge secured their belts and braced for the oncoming impact.

"Mr. Barstow., do we have ion engine power?"

"Yes, sir."

"I want to position the ship so the bottom takes glancing blows and relieves our front from any direct hit."

"It will take about one minute to position ourselves."

"Do we have our magnetic deflectors online?"

"In twenty seconds, sir."

"Under normal maximum operating power when they do, Mr. Barstow. What are our chances of survival?"

"Slim to none, sir."

"I think I can improve on that. Get Mr. Langer on the bridge," I continued to give orders, and in just a minute or so, Mr. Langer appeared.

"Mr. Langer, we need your assistance."

"Me, sir?"

"Yes. We are going to be surfing through territory, and I need your expertise. When you are skiing downhill, how do you deal with ground irregularities?"

"Well, sir, you bend your knees so that your legs become shock absorbers."

"Is there anything else?"

"You shift your weight from side to side so that much of the time, your legs don't bear the full weight of your body."

"Thank you, Mr. Langer. You may return to your station."

I punched in some initiation procedures for the ion engines. "Mr. Barstow, can you arrange to sway the ship back and forth?"

"Not with the ion engines."

"Can we do it with our docking jets?"

"Yes, sir. What rate of shift do we require?"

"Ten-degree shift on both sides over a twenty-second cycle time."

"When do we initiate?"

"Ten seconds before we enter the field. Have we reached the proper angle of entry, Mr. Barstow.?"

"We will reach proper angle in twenty seconds and field entry in forty seconds."

I sat down and buckled up. With my heart in my throat and butterflies in my stomach, I waited to give the countdown.

"Mr. Anderson, on my mark. Start the shift in five, four, three, two, one. Now."

The swaying began, and then the pelting began. We were taking a lot of hard knocks that sounded like popping corn. This continued for two minutes, and I yelled to Mr. Anderson, "Stop the docking jets!"

We stopped our undulating just in time to see a large bolder narrowly miss and hit us head-on. A few more hits, and we were through. Everyone sat for a moment, speechless, and then over the ship's intercom came the sound of elated celebration. Mr. Anderson slowly turned to me and just looked for a moment.

"That was spectacular, Captain."

"Thank you, Mr. Anderson."

We shook hands, and then I reached for the intercom switch. "Mr. Matson. Damage report?"

"Captain, we took a lot of hits, but there is nothing we can't handle."

"What about hull breaches?"

"No critical hull breaches were registered."

"I want a full report when you've finished repairs, Mr. Matson."

I turned to Mr. Barstow, and he was shaking his head. "Captain, I would not believe it if I hadn't seen it."

"Well, I have to do something to justify my place at the helm. Mr. Barstow, I would like you to go help Mr. Matson and find out what is causing all our breech alarms."

"Yes, sir. I'll get right to it."

Again I made an announcement. "Attention, everyone. I want everyone, with the exception of the engineering staff, to go to your quarters, relax, and get some sleep so we will be refreshed when we come back online." I left the bridge and headed to my quarters.

Ms. Cooper and Ensign Clark were coming down the corridor. "Captain, you did it again," said Mr. Clark.

"Thank you. Now take advantage of the downtime and get some rest."

"Yes, sir," they said and continued down the corridor.

I entered my quarters and hoped that I would fall asleep quickly. It was not to be though, for an hour later, I was still awake. I got up, put on some shorts and sweatshirt, and headed to the handball court. When I got there, Ms. Yates was already swatting the ball around.

"Captain Jenner. You got us through another situation."

"That was because of the work of several people, Ms. Yates."

"Yes, Captain. Care for a game?"

"I don't think so. In fact, I'll get us some coffee. I'd like to speak to you about the communications project we have coming up."

"Yes, sir."

"Is everything prepared for the grand experiment?"

"Yes, sir. Every aspect of the experiment has been checked out. But that was before we had the power problems."

"I suggest you do a full diagnostic when we power up and make sure that everything is in order. After all, Ms. Yates, it's your time in the spotlight."

CHAPTER 26

I finally got my chance at the handball court after sending Ms. Yates on her way. Handball was a great way to work out the excess energy you had. With each slap of the ball, you destroyed your enemies— enemies like Billy Lester, the bully that would intimidate me in grammar school, or maybe Ms. Finch, my eighth-grade English teacher who always kept me after class. Wipe them out with one slap of that handball.

It didn't take long before I was physically spent, so I returned to my quarters. Again I could not sleep. I decided to go to the galley for a snack. Flipping through the culinary menu of my mind as I approached the galley, I could see that I was not the only one seeking a tender morsel.

"What have we here?" I said to a startled group of kitchen bandits.

"Captain. Join us," said Ms. Cooper.

"Were you on your way to the galley when I saw you earlier?"

"No, sir. I just couldn't sleep."

"And Ensign Clark?"

"He's here too."

"Oh yes," I said, seeing him at the end of the table. "Look, everybody," said Mr. Clark. "The captain is here."
Everyone applauded and asked me to speak. I could not refuse, so I made it brief.

"I would like to thank you all for your support. If it were not for the contributions that you have all made, I could not function as well in my duties. And I must thank Mr. Langer and his knowledge of downhill skiing that helped me surf through the storm we just encountered. Now I'll shut up, and we can all have a well-deserved snack."

There was more applause, and then everyone settled down to their meal. Our chemists Lorin and Lanie Rodgers were preparing plates for everyone and a more appetizing display I could not imagine. Ms. Dennison approached me with a plate of food in her hand.

"I didn't know you were a late-night snacker, Captain."

"I managed to keep that fact out of my résumé."

"Well, accept this plate I've prepared for you."

"And I see you prepared it with surgical precision."

"Oh, Captain. You are such a flatterer," she said, smiling.

"Is Ms. Grason here?"

"I believe I saw her talking to Mr. Clark."

"Oh yes. I see her. Now I'll be right back, so don't go away."

I strolled over to Ms. Grason and Mr. Clark and found them speaking of our latest adventure.

"Ms. Grason, Ensign Clark. It appears that a number of us had the same idea."

"I just couldn't sleep," said Mr. Clark. "What with all the excitement."

"Shall we be seated?" I asked.

We sat down and continued our conversation. Cal Decker brought a plate and sat down at the table.

"Mr. Decker. You are also a midnight kitchen-raider?" I asked.

"Ever since we got the new cooks. This food is great, and if you ask me, I think we should keep things the way they are."

"I'll take it under advisement, Mr. Decker."

For the next few minutes, we engaged in polite conversation, and then Mr. Decker finished his meal and excused himself. Shortly after, Mr. Clark did the same and left me with Ms. Grason.

"I am very pleased with your work, and without the modifications you made in our robots, we would not be coming back online as quickly."

"Thank you, Captain. I must say that these robots are the best I've ever worked with."

"And with your additions, they have never worked better. You keep up the good work, Ms. Grason."

The Rodgers twins were clearing the table, and I took the opportunity to complement them on their preparations. In their normal, shy, retiring style, they thanked me for the accolades and made a hasty retreat. I returned to Ms. Dennison and found her busily enjoying her repast.

"Like the chow?"

"Yes, this is great. I didn't know we had such great cooks on board."

"I've had several requests to change our culinary practices."

"Count me among them," she said between bites.

"Is that fair to the Rodgers twins?" I asked. "Are you kidding? They love cooking."

"I'll need to check with them before I add this to their job description."

"I wouldn't do that if I were you."

"Why not?"

"They might turn you down."

"They have a right to turn me down."

"You're the captain. You can order them to cook."

"I wouldn't do that."

"That's your problem. No guts."

"What did you say?"

"You heard me."

"I can't use my authority to force a crew member to do something."

"The hell you can't. You're the captain."

"So you think I should rule with an iron hand?"

"Yes. Put them in chains."

"I'll have to check our dungeon and see if we have any chains."

"Don't forget the whips."

"You're a lot meaner than most medical officers."

"Thank you, Captain."

"We don't have a dungeon master. We had to leave port before we filled the position. Would you like to apply for the job?"

"I have much better equipment in sick bay, and I live to torture."

"That sounds very intriguing. I'll schedule you an interview. Just drop by Personnel."

"I have a lot to do in sick bay. We're heavily booked."

"You'll let me know when you have some time."

"First chance I get."

"I didn't know you were so enterprising, Ms. Dennison."

"Got to keep food on the table."

"And cooks in the kitchen, Ms. Dennison? Maybe that's where we need the chains."

CHAPTER 27

After the late-night snack, I returned to my quarters and for the first time in weeks, had a great night's sleep. When I woke up, I showered and dressed quickly for my meeting with Barstow.

When I entered the conference room, Mr. Barstow was already there.

"May I ask what this meeting is about, Mr. Barstow.?"

"We may have difficulty coming back online."

"Why is that?"

"We have lost our clock frequency."

"What is the drive frequency we are using for our present function?"

"A substitute frequency, which is currently in operation, can handle some basic function that we are using now. Only our atomic clock frequency can handle the speed required for full function of a starship, and we have lost our link."

"And how did that happen?"

"Our quantum link was severed."

"I was not aware that it was possible."

"Nor was I, Captain."

"What can we do about it?"

"A rather unusual procedure of my own creation."

"What would have inspired you to develop such a procedure?"

"You never know what you will encounter in deep space."

"What does the procedure require?"

"We need five terminals, and we should start the process immediately."

I pressed the ship's intercom and made the announcement. "Attention, everyone. Report to the conference room for a briefing." Ms. Yates was the first to arrive, and she asked if our meeting concerned the communications project. I assured her that it did not, and she took a seat. Mr. Clark was next, followed by Ms. Kane and Ms. Cooper. The remainder of the crew showed up in the next few minutes, and the meeting started.

"We have another situation that has arisen. I will turn the floor over to Mr. Barstow to explain."

"I must first provide you with a bit of background. All computer systems require a clock frequency to operate. Most computers have their own drive clock built in. Starships require a frequency that can only be supplied by an atomic clock. The atomic clock derives its frequency from subatomic activity and is linked to all terminals on the ship. Our clock interface is not recognizing the atomic clock frequency."

"Is something wrong with the clock interface?" asked Mr. Decker.

"No, and the only way to test the device is to do an isolated interface boot up using five terminals. We cannot produce a frequency equal to the atomic clock. We can produce a signal that will operate the computers at a greatly reduced speed. The problem occurs when we try to switch frequencies. The only way to accomplish this is to open just five terminals and run the five programming cards in front of you. In addition to the bridge terminal, we will require the participation of terminal 3, 7, 9, and 10. I will come to each terminal and give written instructions on what you are to do. This will take place right after this meeting. Are there any questions?"

No one spoke up, so I adjourned the meeting. After all had left, I sat down with Barstow and asked him to given me an assessment of the procedure. He was not sure of the outcome, and although he seemed confident, I could see that he was concerned about the procedure. I resisted the urge to ask if there was an alternate procedure for I suspected there wasn't.

The next few minutes seemed eternal as I waited for him to return to the bridge after giving individual instructions to the other four stations. Everyone was anxious to get back to normal and get back to the mission, so the next few minutes were critical. Mr. Barstow returned to the bridge, and I waited for him to start the process. He booted up the synthesized drive frequency, and the screen lit up. He continued to enter data and signaled for me to start.

"I have entered the transfer code," he said. "Interlock transfer engaged. Station 3," he announced, "enter your program now." More data was entered, and then he gave the next announcement. "Station 7, run your program now." He pressed the key, and the screens went black. It didn't work, so Barstow gave the order to reset, and the next try gave the same results. Mr. Barstow thought for a moment and then started entering data at a furious pace. He then announced, "All stations, place the program cards in your terminal. I will run sequencing from the bridge starting now. Is there any response on station 3?"

"Yes, sir. It just came up on station 3."

"Station 7?"

"Yes, we are functional."

"Station 9?"

For a moment, there was no response, and then..."Station 9, working."

"Station 10?"

"Station 10, functional."

I could hear cheering all around the ship, and suddenly the screens blank, followed by moans and groans.

"No, no. That's supposed to happen," said Barstow. "In thirty seconds, the entire ship will reboot."

For a second time, cheers could be heard while the ship started

113

coming to life. We were back online. I turned to Mr. Barstow and saw the closest thing to a smile that I've ever seen on his face. "Congratulations, Mr. Barstow, for an excellent job."

"Thank you, sir," he said.

CHAPTER 28

In just moments, the crew was filing in with unabashed enthusiasm. "Shall we address your fans?" I asked.

"By all means," he replied with a restored self-confidence. I was certain that Mr. Barstow felt responsible for our problems and obligated to solve them. And solve them he did, in quite a dramatic fashion.

An appreciative crew was anxious to congratulate Barstow for a job well done. The Rodgers twins even baked a cake with "Thank you, Science Officer Barstow" written on it. Everyone was up for another party, and I must admit, so was I. We opened several bottles of our best vintage and poured liberally. Barstow was looking pleased, and I took the opportunity to make an announcement. "I would like to propose a toast to our ingenious science officer, Mr. Barstow." The customary cheers ensued, and then came time for party banter and cake. Barstow was enjoying the attention, so I left him to get a taste of the treat provided by our master chefs. As I started to slice a piece, Ms. Dennison approached with a plate.

"Cut me a slice of that," she said.

"It does look good. Does look good."

"Bigger than that," she said emphatically. "I didn't know you were such a pig."

"Check my résumé."

I sliced a large piece, and she, like a kid, dug in. Unable to speak with a mouthful of food, I took the time to get the upper hand.

"So is it good? Oh, I'm sorry. You shouldn't talk with your mouth full."

She tried to swallow quickly and ended up with cake on her chin. "You have cake on your chin," I said like a father would say to a small child.

"I'll get you for this, Captain."

"Here, let me get some you missed. Or maybe I should wait until you are finished and then hose you down."

"You know, pigs don't like to be hosed down."

"That's right. What could I have been thinking?"

"What indeed? This cake is great. I told you to keep the Rodgers twins locked in the kitchen."

"That's galley, Ms. Dennison."

"Galley, kitchen—who cares? They should be chained to a stove."

"That's oven, Ms. Dennison. And what is this preoccupation with chains?"

"That's also in my résumé. Before I was a medical officer, I was a dominatrix."

"Oh, really. How interesting."

"Yes, it really prepares one for a medical career."

Ms. Grason came over with Archimedes at her side. The robot was producing audio signals that were pleasant and conversational.

"Captain, I thought I would bring Archimedes to the party."

"By all means, Ms. Grason. He performed exceptionally and should be a part of the celebration."

"Thank you, Captain," she said like a proud parent.

"Excuse me, Ms. Grason," I said. "I must speak to Ms. Cooper."

I stepped over to the serving table where she was cutting a piece of cake.

"The cake is delicious."

"Yes, I was told."

"May I pour you a glass of wine?"

"Yes, Captain."

I poured her a glass and felt something nudge me on my leg. "Archie," I said. "Would you like a glass of wine?"

"Archie? Did I hear you say Archie?"

"Yes, Ms. Cooper. Archie, short for Archimedes."

"Cute," she said, rolling her eyes. "Is something wrong, Ms. Cooper?"

"I guess I'm not in a party mood."

"And why is that?"

"I miss home so much."

"We all miss home."

"Lately, we have had a lot of time to think about things."

"And you were reminiscing?"

"Yes, I guess I was."

"Go ahead. Try the cake."

She took a bite and almost cried. I reached for a napkin and handed it to her.

"Ms. Cooper, what's wrong?"

"It tastes like my grandmother's cake."

I took the plate from her and helped her to a seat. I looked across the room and saw Ms. Dennison. She saw me motioning, and she came over.

"What's happening?" she said.

"Oh, I think we have a case of homesickness. I thought you might be able to help."

"How, by singing some nostalgic tunes?"

"I just thought you might have something for her."

"I can't cure intoxication."

"She's drunk?" I asked. "As the proverbial skunk."

Just then, Ms. Cooper passed out and fell forward on the table. saw Mr. Clark coming our way. "Mr. Clark," I said. "Could you help Ms. Dennison take Ms. Cooper to her quarters?"

"Yes, sir," he said. "I told her not to overdo it."

"I'll be back," said Dennison. "I haven't finished my cake."

I poured another glass of wine and sat down. The party was still in full force when suddenly we heard something impact the ship. Quickly, everyone fell to the floor. Again we heard another impact and then a third. I could not believe that we were in for another pelting.

CHAPTER 29

We were scurrying about in panic, not knowing what to do. I jumped and ran to the bridge. Being tipsy, it took more effort than I had anticipated. Finally though, I set foot on the bridge.

"What are we being hit by, Mr. Anderson?" I asked.

"Nothing I can see, sir."

"Are our sensors online?"

"Online and functioning, sir."

"Then why can't we see what is hitting us?"

"Because nothing is, Captain," said Mr. Barstow, approaching from behind.

"Explain, Mr. Barstow."

"If you will look at the screen, sir."

Before my eyes was a magnificent display of light, all colors imaginable in a flashing light show of incredible beauty.

"The light display is caused by our light source encountering a plasma field. The color changes occur when there is a power transfer. Like a lightning storm, the sounds are created when these power transfers interact with the surface of our ship."

"So it is a sound and light show," I stated. "You could say that, sir."

I pressed the intercom button. "Attention, everyone. Report to the bridge on the double."

In just a few minutes, they filed in and were awed by the fascinating show. Like in a darkened theater, they found their seats while keeping their eyes on the screen. For a few minutes, they were mesmerized by the incredible display. I walked in front of the display and halted the show. "The show you have just witnessed is the result of a plasma field coming in contact with our light-source destination. Ladies and gentlemen, we have arrived."

The announcement was greeted with applause and what seemed to be a form of group relief.

"It is time to get back to work and find out what we are dealing with, so if everyone will return to your stations, we can get on with it."

I asked Mr. Barstow to join me in the conference room. "I'll be there in just a couple of minutes, Captain."

"Would you care for coffee?" I asked.

"Yes, sir."

"I sure the hell would," I said under my breath.

I started coffee and sat for a minute, observing the light show on a computer monitor. It felt good to have our power back and be in control once again. I sat back and took a sip of coffee. Mr. Barstow returned and took a sip of his coffee. "You wanted to speak to me, sir?"

"Yes, I wanted to clear up a few things."

"What can I clear up for you?"

"Well, I noticed that the sound accompaniment to the light show has diminished in amplitude considerably."

"I made an adjustment in the ship's dampening system."

"We have a sonic dampening system?"

"Yes, sir. It's part of the IGS system."

"You learn something every day."

"I can reduce it further if you like."

"No. It's kind of pleasant, like rain."

"I will leave it where it is."

"Now that everything is back to normal, I would like an update on the progress of research in each department, and if you'll inform all concerned that I expect reports in twenty-four hours in the conference room."

"Yes, sir."

"One final thing, Mr. Barstow."

"What's that, sir?"

"I would like to know what your assessment of the mission is."

"We have had major setbacks, but we have managed to recover in time to do what we were sent here to do."

"Any predications?"

"I prefer not to speculate."

"If you do feel in a speculative mood, you'll let me know?"

"The moment I feel the mood."

"Thank you, Mr. Barstow. You may return to duty."

"Yes, Captain," he said and then left the room.

I stayed for a few minutes and savored the moment, and I didn't want to lose the buzz I had going. We were back on track, and I felt that we could handle anything. After about twenty minutes, I had another cup of coffee to steady myself for the walk back to my quarters.

On my way back, I caught myself humming a tune and as if outside myself, questioned why I would do such a thing. Ms. Kane interrupted my self-analysis at my door.

"I must speak to you, Captain," she said with tension in her voice.

"What about, Ms. Kane?"

"It is about the research you asked me to do."

"Can't it wait until our scheduled meeting?"

"No, sir," she said.

"Very well. Let's go to the conference room."

"No, sir. We need to go to hydroponics."

"Very well," I said as she continued. She looked frightened and unsteady on her feet.

"The readings of our atmospheric instruments were correct."

"Ms. Kane. We have come through a series of unusual and

phenomenal events, any one of which could have had a major impact on our instrumentation. There is no way to know what event or events accounted for such inaccuracies. During the time we were laboring under these conditions, nobody was even breathing hard. Our air-quality analyzers measure the percentage ratio of composite gasses compared to what is normal for us. If those readings were anywhere near accurate, we'd all be dead."

"The fact that our portable units read the same should be of concern to you," she insisted.

"Ms. Kane, I can't account for that, and maybe we never will. But there is no way the readings were accurate."

"Captain, I ran an experiment of my own in hydroponics that was independent of the ship's instrumentation."

The hydroponics lab was just ahead. I pushed the door button, and it slid open. She stood there for a moment as if she were afraid to enter.

"Go ahead, Ms. Kane."

She stepped into the lab and said, "It's over there."

"What's over there?"

"The hybrid herbs."

"What's that?"

"It's a project that I was working on before I became a crew member of the Mobius."

"Oh. It's you, Captain and Ms. Kane," said Mr. Greyson as he approached. "What can I do for you?"

"We just came down to take a look at Ms. Kane's project."

"Oh yes. It's over here."

We stepped over to the station with computer terminal. Ms. Kane sat down and started entering data.

"Captain, the herbs you see have been genetically engineered to resist a number of adverse conditions. Among them is the ability to survive a rather severely altered atmosphere ratio. Look at these plants, Captain. They are all dead."

"And what do you think killed them, Ms. Kane?" I asked. "First, let me ask you a question, Captain."

"Yes?"

"Are we back online 100 percent functional?"

"Yes, Ms. Kane."

"And you would have complete confidence in any measurement taken by the system in its present condition?"

"Yes, Ms. Kane, I would."

She placed the plants on an analyzing tray and inserted it into the examination slot. After she programmed the unit and in a few moments, the screen lit up with information. A full description of the type of plant and a chemical breakdown of the plants was scrolled on the screen. The plants were identified as dead due to a reapportionment of the ratio of gas content in the atmosphere. The maximum ratio differential was estimated to be 60 percent.

CHAPTER 30

"That's exactly what the ship's analyzers read!" she yelled. I could see she was trembling, and so I came up with something that I felt would calm her down.

"I'm going to ask our science officer to work with you to find out what we are dealing with. And you, Mr. Greyson. You should be a part of the research team also."

"Yes, sir," said Mr. Greyson, sounding pleased at my decision. "I will send Mr. Barstow down as soon as I return to the bridge."

She seemed greatly relieved to know that Barstow would be working with her. This seemed to be a priority situation, so I decided to make it happen immediately and hastily made it to the bridge. When I got there, Barstow was not present. I asked Mr. Anderson where Barstow was.

"He's in the conference room, sir."

I found him taking a caffeine break. "Mr. Barstow. When you have finished your coffee, I would like you to report to the hydroponics lab and help Ms. Kane and Mr. Greyson with an important project."

"Yes, sir. I expected this."

"You did?"

"Yes, sir. I figured that something odd would show up somewhere."

"Well, get down there soon. Ms. Kane and Mr. Greyson need your help."

"I'm on it, Captain," he said as he turned and walked out the door.

I noticed Ms. Cooper standing at the door, and I asked her in.

"Oh, Captain. I want to apologize for my behavior at the party."

"You don't need to apologize, Ms. Cooper."

"I'm not much of a drinker, and I just had too much."

"I understand."

"I'm so embarrassed."

"Now, Ms. Cooper. You shouldn't feel embarrassed."

"I don't know what came over me."

"You were just homesick, and you had a little too much to drink."

I put my arm around her shoulder and tried to cheer her up, but she just wouldn't stop apologizing. I then thought to take a more formal approach.

"Stand at attention, Ms. Cooper," I said in my most authoritative manner.

"Yes, sir," she answered with surprise.

"Now I don't want to hear any more about this, Ms. Cooper. I want you to turn around and get back to your station on the double."

She almost ran out. This is great. I'll have to keep this in mind for those awkward situations. Maybe it was time that I acted with more authority—more formal, more military, more stuffy, more pomp- ous…maybe not.

The next few days were relatively uneventful and a welcome change from the recent chaos. While our science personnel were busy at work, I took the opportunity to get back to my golf game. Mr. Brock, a golfer himself, designed his own set of golf clubs and volunteered to make me a set. We were both looking to play eighteen holes, so we decided to make it a twosome. He asked me if I was related to last year's PGA winner, Frank Jenner. Yes, I told him. He's my cousin. Mr. Brock was impressed, and then I told him I was only an average golfer. He assured me that the clubs he'd make for me would change all that. I was delighted to hear it even though I didn't believe it. We played a few holes while he analyzed my style and I played my typical game.

He said he could have them ready for me in two days. I was pleased to hear that, and sure enough, they were there in two days. I could try them in private, and when I took out the driver, it was the weirdest-looking thing I had ever seen. The shaft seemed bent in several places, and it seemed to be too light to have much impact. I tried it anyway and hit a shot over three hundred yards that landed me on the green just four feet from the hole. A short put and I was under par. These clubs were fantastic. They took ten strokes of my score. I couldn't wait to play someone. That would have to wait though for I had a lot of reading to do regarding the next phase of our mission.

Our current state-of-the-art quantum physics enabled us to harness the potential power of subatomic particles. The early years of quantum physics were, to say the least, tedious. The most significant barrier to progress was the Heisenberg uncertainty principle. Not until Dr. Elton Harcourt proved that despite the uncertainty principle, control of subatomic particles was attainable. His technique of precisely separating quarks from their host led the way to further control techniques.

My friend Paul Stockton was a major contributor to the development of many quantum innovations. Deep-space missions had been successful in collecting new substances that contained new types of particles. The breakthrough, though, came with the development of a quantum control matrix. Using the characteristics of one type of particle to control the actions of another was the key to the success of the matrix. Because building a matrix involved growing containment walls, the term quantum farming was born. To coax specific particles to assume a control position, a seeding plane was fabricated to act as a particle allure field. High-frequency electromagnetic stimulation could achieve a number of controlled effects. The exact outcome depended on the nature of the seeding plane, the type of injected particle, and the frequency of the electromagnetic control field.

The creation of other substrate seeding fields for the manufacture of power generation units was just one of the uses of the technology. Separating entangled pairs was the first step in the development of the parallel effect syndrome, also known as the Corsican effect. When an electromagnetic signal is applied to one of an entangled pair, a reaction of equal intensity can be observed in the other particle regardless of proximity. Separated pairs had the potential of being used as signal processors to provide instantaneous communication across vast areas of space. Only when we learned how to create containment chambers did the process actually become practical.

CHAPTER 31

Ms. Yates was concerned about the procedure since she was the communications officer. She asked for a special briefing with Mr. Barstow and myself. She was on time and nervous as she entered the conference room. Barstow had set up all the information codes regarding the procedure on our main viewing screen.

"We must decrease our speed before the process can be initiated," he said. "Our course must be maintained for five hours."

"And our duties take place during the five hours?" she asked.

"Yes," said Barstow. "In an exact ordered sequence."

He went to the console and entered data, and the screen lit up with a procedure list and system flow chart. Ms. Yates looked at the screen intensely as I put a cup of coffee in front of her. She reached for the cup and fainted. Mr. Barstow called on the ship's intercom for medical assistance. I was on one knee trying to revive Ms. Yates when Ms. Dennison stepped in. I told her that Ms. Yates had just fainted, and she started her examination. After a minute or so, Ms. Yates regained consciousness, and we helped her to sick bay.

"I'm okay. I don't need to go to sick bay."

"Now, Ms. Yates, you just fainted. You need to be checked out," I said.

"But the procedure. We've got to get back."

"Ms. Yates, the procedure can wait."

After we managed to get her to sick bay, Ms. Dennison gave her a sedative, and she was resting calmly.

"What's wrong with Ms. Yates?" I asked.

"I don't know, but there are three other patients exhibiting the same symptoms."

"And who would those be, Ms. Dennison?"

"Mr. Palasco, Mr. Greyson, and Ms. Kane."

"Ms. Kane and Mr. Greyson were working on a project in hydroponics lab, but Mr. Palasco…"

"Oh, Captain. Mr. Palasco has his repair shop right next door."

"With Ms. Kane missing, you must need some help."

"You read my mind, Captain."

"I'll get Ms. Grason to help. She was once a medical assistant."

She thanked me, and I started to leave. But I paused, and I voiced an afterthought.

"When you get this situation under control, give me a buzz. I want to know what you think of Ms. Kane's research."

I was getting an uneasy feeling again. We had just come through several stressful situations and were beginning to relax. This crew needed a break. I headed back to the bridge to speak to Barstow and met him in the corridor.

"Mr. Barstow. I would like to speak to you about your work with Ms. Kane."

"Yes, sir."

"This may take a while, so if you start some coffee, I'll join you in the conference room as soon as I check with Mr. Anderson."

When we got to the bridge, I could see the continual light show on our main screen. Mr. Anderson was looking at it as if in a trance. I stood there for a few minutes, and not once did he look around.

"Mr. Anderson," I said with authority. "Yes, sir," he said, startled.

"Put a filter on that screen."

"Yes, sir. I'm sorry, sir."

"Just don't get caught up in it."

"Yes, sir. I mean, no, sir."

"Is there anything I should be aware of?"

"No, sir. All is normal. Well, you know what I mean."

"I think I do, Mr. Anderson. Stay alert."

I retired to the conference room and saw Barstow preparing his presentation. I sat down and took a sip of coffee while he continued programming. I was hoping he'd keep working for a few minutes and give me time to catch my breath. As if he read my mind, he did take a few minutes to finish his programming. I even had a chance to lean back and close my eyes for a few seconds. I could hear that Barstow had finished, so I sat up just as he turned to face me. He did not rush into a dissertation. Instead, he sat and took a sip of coffee.

"I'm a little tired, Mr. Barstow," I said with a sigh. "Yes, sir. I understand," he said sympathetically. "What would I do without you, Mr. Barstow.?"

"I have no idea, sir."

"Well, do you have some information for me?"

"Well, sir, my collaboration with Ms. Kane has revealed nothing that can be considered unusual or abnormal."

"A variety of maladies have broken out, and everything is normal?" I asked.

"I didn't say that. I said that nothing appears abnormal."

"What do you think? And don't tell me that you want to avoid speculation?"

"We have some kind of virus."

"We have what?"

"A virus that defies detection by our medical analyzers."

"How can that be?"

"Our analyzers are state-of-the-art. They can detect anything. All of the symptoms suggest a virus. We just haven't isolated it."

"How is it transmitted?"

"It's airborne."

"Then we are all susceptible."

"Yes, sir."

"Until we find out what this is, I'm assigning you to sick-bay duty with Ms. Dennison and Ms. Grason."

"Yes, sir."

CHAPTER 32

My mind was racing with doomsday scenarios, and I could not get my mind on anything else. I returned to my quarters and tried to relax. Maybe reading would get my mind off my problems. What would it be: history, biographies, detective stories, or science fiction? No, I hate science fiction. All those outer-space monsters are so ridiculous. Give me a good old-fashioned detective story.

I looked over the book list and saw one of the new series of stories by Charles Masters. I picked it up from the shelf and started reading.

She stepped into my office, a breathtaking beauty. Crying her eyes out, she explained that her husband was missing. I tried to console her, but she really had the waterworks turned on. She sat down and crossed her legs. I offered her a drink to calm down, and in a few minutes, her sobbing subsided. I gave her a tissue to dry her eyes, and a tear dropped on her breast and down her cleavage.

I skipped a few pages and started reading again.

I kissed her hard, and she melted like butter in a hot pan. I skipped a few more pages.

Her breasts struggled to free themselves from their chiffon bondage.

I put the book down and wondered why it was on the best seller list. Maybe it was his versatility. Detective services with a free breast examination.

I had to look elsewhere for my distraction. Mr. Anderson was enthralled by our light show, so I gave it a try. I punched it up on my big screen. At first, it was just a relaxing montage of interesting shapes and vivid colors. As I sat there, the shapes became more and more mesmerizing and intermingled with memory. After falling asleep, my dreams became an extension of the experience, and it was very therapeutic.

I awakened refreshed and ready to face the day. While getting dressed, I found myself humming and decided to recommend this to all crew members. My personal communicator buzzed, and I tapped it on.

"Captain, this is Barstow. We have discovered a derelict ship just ahead, and we need to investigate."

"Meet me in docking bay 1, Mr. Barstow."

I immediately headed to docking bay 1. Mr. Barstow was already there, and we both got into one of the small shuttles. I booted the ship's computer and read off the checklist.

"Communications, Mr. Barstow?"

"Check."

"Laser guidance?"

"Check."

"Proximity sensors?"

"Check."

"CCP system?"

"Check."

"Looks like we're ready to go, Mr. Barstow."

"Ready, sir."

"Bay doors open."

"Doors open, Captain."

"Mr. Barstow, engage."

We slowly lifted off and headed out. It felt great to pilot one of these runabouts once again. As soon as we cleared the bay doors, I powered up, and we shot out like the proverbial "bat out of hell."

"I didn't know you were such a hot-rod racer," said Barstow. "If I weren't captain of the Mobius, I would be on the racing circuit."

"Is that a fact, sir?" he said in disbelief.

I gave him a demonstration of what I could do, and he reminded

133

me that we were off course.

"Well, punch it up, Mr. Barstow," I said. "And we'll be on our way."

"Yes, Captain."

"In all our haste to get going, I forgot to ask what we were looking for."

"I'm not sure."

"How big is it?"

"Larger than the Mobius."

"Have we seen anything like this before?"

"No, sir. It is some kind of alien ship, and at our present rate of speed, we should be in visual range in about thirty seconds."

CHAPTER 33

Just ahead, we saw something coming into view. I slowed the ship as we started closing in. I turned on our external lights, and the oddest- looking structure appeared before us. I started a life-form scan and slowly circled the vessel. It looked crusty, as if covered with barnacles. Strange markings and configurations covered the surface, and definite signs of instrumentation were in evidence.

"Mr. Anderson," I said over the ship's communication. "Engage tractor beam."

"Yes, sir," he replied.

We continued our search around the gigantic ship as we looked for possible access. The awesome sight looming before us was both disconcerting and riveting. Remnants of an advanced and ancient civilization were exhibited before our eyes as if we were on some galactic museum.

"Just ahead, Captain," said Barstow. "A large hole."

"Can we fit?" I asked.

"It will be close, but I'm sure you'll manage, Captain," he said, as if to reassure me of his confidence in my ability.

I maneuvered the shuttle closer to the hole, and it seemed to be just big enough to enter. As we closed in, Barstow made an observation. "It's a warship."

"A what?" I said.

"A warship. The instrumentation suggests weaponry and this hole was blasted."

"Very interesting."

"Even more interesting, I'm getting a positive life-force reading."

"How can that be, Mr. Barstow?"

"I can't say, Captain."

The eerie innards of the ship looked more organic than mechanical. Everything was crusty and corroded, a cavernous mangle of broken and bent structures.

"What's that?" said Barstow.

"Where?"

"To the left."

I turned the ship to the area Barstow indicated. As we approached, a few objects came into view. At first, they looked like more debris, then a closer look led to a frightening discovery: hundreds of petrified alien bodies everywhere.

"Oh my," I said. "Will you look at that?"

"Yes, Captain," replied Barstow. "What a sight."

"I think I'll try to get a closer look."

I maneuvered the ship close to one of the bodies. Just like the ship, it was crusty and covered with something corrosive.

"Activate the manipulators," I said.

Mr. Barstow started the activation procedure for the frontal manipulators. I jockeyed for position, and Mr. Barstow took his seat at the manipulator console. He used the left manipulator to lift the body and move it for examination. The viewer was set into place, and a magnification of ten was selected. The screen revealed a hard bluish coating over the body. A small piece was broken off with the micro-manipulator and placed on a small retainer platform. Close examination of the sample showed that what seemed to be an external coating of blue material was actually a substance produced by the organism itself.

The creatures were bipeds of approximately seven feet tall and had a tail. Their heads were a bone-like material with two eyes and an upper jaw with separated teeth that fit into individual holes in the lower jaw. The top of the head was formed into a sharp ax-like shape that could be used as a weapon. Their hands and feet had five digits with an opposable thumb and long claw-like nails. Elbows, knees, and heels also had sharp bone structures that were also sharp weapons.

"They are survivors," said Barstow. "Hard to imagine anything that could defeat them."

"Obviously something did."

"I would like to take this back for more complete analysis. I believe we have enough space in the undercarriage sample bin."

"Very well, Mr. Barstow."

He slowly slid the sample into the bin and closed it tight. The manipulators were retracted, and we were free to go.

"Now we can go and investigate that life-form reading," I said. I turned the shuttle around and followed the tracking signal.

It led us to a large corridor somewhat free of debris, which gave us an opportunity to see more of the instrumentation. A soft glow from an unknown source seemed to indicate something operational on the ship. Blue riblike braces made up most of the wall surface with occasional panels of strange instrumentation. The signal grew stronger as we continued.

CHAPTER 34

Just ahead was a circular hatch. I reactivated the manipulators and positioned them to pry the hatch open. It was difficult to move, so I applied a torch. After a few moments, the hatch began to loosen, and then it opened fully.

A mist hung in the air, which quickly cleared. For a moment, there did not appear to be anything there, and then we were attacked by small creatures. They attached themselves to the shuttle and oozed a slimy substance.

"What is that?" I said.

"I can't be sure, Captain, but they might respond to a high-voltage power charge."

He pressed the button on the console, and they jumped away. But they came right back. I backed the shuttle out slowly for I could not see.

"Mr. Anderson," I asked. "Guide us back to the ship from the bridge."

"Yes, sir," he replied.

For what seemed like hours, we were slowly extracted—The whole time, trying to shake off our unwelcome guests. The front-view port was covered in the ooze, and it became an extremely uncomfortable trip. Our attempts to dislodge our parasites were having little effect.   "Do you find this as unnerving as I do, Mr. Barstow?"

"Probably not. But I have a scientific interest."

"We are about to execute docking maneuvers, Captain," said Mr. Anderson over the ship's communication.

"We are ready, Mr. Anderson," I replied.

"Captain, set your docking interface control to external."

"Done, Mr. Anderson."

The shuttle was slowly drawn into the ship. Magnetic alignment was achieved, and the docking clamps were engaged. I started to power down as the bay doors closed and waited for bridge instructions. While sitting there, I could see the creatures falling from the ship. In a few minutes, they were all on the docking bay floor.

"Now what do you think of that?" I asked Mr. Barstow. "Heat," he said.

"That's it, Mr. Barstow?"

"Yes, sir. These creatures' metabolism is geared to extreme cold."

"And as the temperature increases, a comatose state is brought on?"

"Very good, Captain."

"It's the company I keep," I replied.

"Isolation procedure engaged,Captain," Mr. Anderson announced.

The access tube lowered, and an air lock was achieved. Mr. Barstow and I climbed up the ladder and entered engineering.

"I want to take a look at the creatures," I said to Mr. Matson.

He turned on the viewing screen, and we saw the creatures on the floor. None of them were moving. In fact, they looked dead.

"Now what were you saying about heat, Mr. Barstow" I asked. "These creatures function only in subzero temperatures. They are not a carbon-based life-form."

"And what were they doing on board that ship?"

"They were bred for food. At room temperature, they are totally incapacitated."

"How could a ship that old maintain room temperature?" I asked.

"The ship is not old at all."

"How old do you think it is?"

"I can't be certain, Captain."

"Give me an educated guess, Mr. Barstow."

"The battle that this ship lost could have taken place within the last month."

"You mean this thing was fighting within the last month?"

"Yes, Captain."

I couldn't believe what I was just told. It took a minute to gather my thoughts. We had just come in contact with alien life, and it was a lot to digest.

"I want all medical and research personnel to report to engineering to see what we brought back with us. Make the announcement, Mr. Barstow," I said.

While I waited for the crew members to arrive, I programmed the console for several demonstrations. The creatures looked dead as they remained motionless on the floor. Rusty-colored, flat, and round like pancakes, they were capable of leaping fifty feet or more. No perceivable eyes and no air intake made these among the strangest species ever encountered. Like sharkskin on one side and slimy on the other, their method of survival defied speculation. The staff started filling in as we completed the programming.

"As you all know by now," I said, "we discovered an alien ship, and we shuttled out to it and brought back samples—samples that I'm sure that you all will be interested in examining. If you will step over to the docking bay window, we have something for you to see."

They took their places and waited for the demonstration to begin. "The derelict ship we found is larger than the Mobius and appeared to be a warship. It was badly damaged from battle. Inside we found bodies everywhere. We brought one back for examination, and we also brought something else back. If you will look at the docking bay floor, you will see some disc-shaped organisms. They have no discernible features and go into a dormant state as temperatures rise."

"How can any animal function at those temperatures?" asked Ms. Grason.

"Mr. Barstow. Would you please answer the question?" I said as I relinquished the floor to him.

"Yes, sir. There are several species on Earth that can function at extreme temperatures. For example, organisms on Earth have been found in geysers that can survive in water near boiling temperature. Some fish have been discovered at great depths under great pressure.

The type of life we discovered is silicon-based and functions at extremely low temperatures."

He stepped to the console, made some entries, and returned. "I have lowered the temperature in the docking bay, and in a few minutes, you will witness an amazing resurrection. Are there any questions?"

"Yes, sir," said Ms. Kane. "What are these creatures doing on board a warship?"

"They served as food for the larger species."

"There is something happening in the docking bay," said Ms. Grason.

We all looked up and saw some movement. The creatures started moving and flopping around. They became more and more active and then were jumping around. One of them was flying around like a Frisbee and landed on the window. Others followed until the window was totally covered.

CHAPTER 35

Everyone was startled as the slime started dripping down the window. "What is that?" asked the Rodgers twins in unison.

"It is something that you will be working on," I said.

Mr. Barstow started warming up the docking bay, and the creatures fell off the window to the floor.

"I want a full report on the chemical makeup of these creatures as well as the thing you are about to see."

I turned the shuttle around on the docking turret and opened the storage bin. I could hear the reaction when the alien came into view.

"Good lord. How big is that thing?" asked Ms. Cooper. "Just over seven feet," replied Mr. Barstow.

"It looks like it's covered with something," said Ms. Cooper. "We think it's a secretion from the creature itself."

We loaded the specimen into a transfer unit and sent it to the isolation chamber in lab 2. I stepped forward and made an announcement. "I want all lab personnel to give this your undivided attention. I'll expect a report from each of you in a meeting which is to take place in the conference room forty-eight hours from now. Everyone is dismissed."

They filed out of engineering, and I had a chance to speak to Barstow. "I would appreciate your helping all of our staff by offering them the benefit of your expertise, and maybe the collaboration will give us a more accurate picture."

"Yes, sir. I will get right on it," he said as he left engineering.

I returned to the bridge to speak to Mr. Anderson about his excellent handling of the retrieval and remote docking procedure.

"I would like to commend you on the way you executed the maneuvers in a tight situation."

"Thank you, sir," he said with a degree of pride in his voice.

"Your navigating skills proved to be very useful."

"I'm glad to be of service, Captain."

"Are we back on course?"

"Yes, sir. Steady as she goes."

"Do you have anything to report, Mr. Anderson?"

"Yes, sir. I was doing a scan of our underside as a routine procedure, and I discovered something interesting."

"And what is that, Mr. Anderson?"

"We have no scratching or scaring of our underside despite the pelting we took."

"Do a scan now. Let's take a look."

He programmed an underside camera scan, and as he said, not a single scratch appeared on the ship's surface.

"Sir, we should have all kinds of surface scratching or scaring from our trip through the asteroids."

"Mr. Barstow is occupied for the next forty-eight hours, but when he is free, I want you to show him what you just showed me."

"Yes, Captain."

"And keep me informed of any other oddities you may encounter."

"Will do, sir," he said with confidence. "I will be at lab 2 if you need me."

I was anxious to see what was going on in lab 2 as I walked down the corridor. The entire team of scientists was so hard at work that they didn't even notice my entry. Mr. Barstow was working on the main console with Ms. Kane.

"How's it going, Mr. Barstow?" I asked.

"Oh, Captain, I didn't notice you were here," he said.

"That's quite all right, Mr. Barstow. Do you mind if I watch for a while?"

"Not at all, Captain. If you'll take a seat, I'll explain what we are about to do."

I sat down and waited for Mr. Barstow to start. The instrumentation was impressive. Never had I seen it all on display like that. Multiple microscopic cameras, robotic probes, laser scalpels, and other numerous processing units were all set up, ready to use. I felt privileged to observe our staff doing such important work. Barstow made some computer entries, and the body of the alien came out on an examining platform inside the isolation chamber. He entered more computer codes, and the cameras positioned themselves for scanning.

"We are preparing for our first pass, Captain."

He started the procedure and looked up at the monitor. The first pass was without magnification, and it was a nauseating sight.

"I'm going to take a small sample for high-power examination."

The laser unit moved into place, cut a small piece, and placed it on a viewing slide. It was then transferred to the electron microscope where Ms. Kane was setting it up for viewing. Barstow made another cut and peeled back a layer.

"You can see the hard outer layer is actually a secretion of the alien's body. Small vessels over the entire skin surface excrete this fluid that hardens and protects the body from injury. In the event of multiple injuries, the entire body covers itself to stop bleeding and prevent contamination."

"Built-in bandages," I said

"Very good, Captain," replied Barstow. "Makes them well suited to be warriors."

"Right again, Captain."

"What is the chemical breakdown of the substance?" I asked.

"The creature is silicon-based, so you might say it's a form of glass."

"Can we estimate their restorative time frame?"

"I can only guess, but I would say anywhere from ten to one hundred times faster than our own recuperative abilities…depending on temperature."

"And how does that work out?" I asked.

"They are comfortable in the same temperature frame that we are. The temperature in the examination chamber is five degrees Celsius to retard cell deterioration as we perform our examination."

"There is no perceivable heart or lung configuration," said Ms. Kane.

"What is their place of origin?" I asked.

"Well," said Ms. Kane. "Let's just say…they're not from around here."

It was good to see that Ms. Kane still had a sense of humor. Mr. Barstow turned to make some entries on the computer and then turned back to face me and resume our conversation.

"What are they doing here?" I asked.

He looked surprised at the question and paused for a few moments.

CHAPTER 36

I decided to continue our conversation in the conference room, so when we sat down at the big table, I asked again.

"What are they doing here?"

"They are probably here for the same reason we are: to investigate the light…oh yes, and to look for a cold planet."

"For what purpose?" I asked. "Farming."

"Did I hear you correctly? Farming?"

"Yes, Captain."

"Farming what?"

"Those pancake-shaped creatures they use for food."

"Are you serious?"

"Yes, sir. And they found it."

"What do you mean they found it?"

"Yes, sir. The planet Neptune."

"How do you know this, Mr. Barstow?"

"When we were on the shuttle, I noticed a faint life-form reading coming from Neptune. The alien ship was low in their stock of food, and the creatures only breed in a frigid atmosphere. So Neptune is ideal."

I pressed my personal communicator and asked for the bridge. "Mr. Anderson. Would you send out a probe to scan the surface of Neptune?"

"Yes, sir. Right away."

I got up from my chair to get a cup of coffee. I asked Barstow if he wanted one, and he nodded. So I came back to the table with the two cups.

"Continue your description of our main guest."

"Yes, sir," he said and then stepped to the control panel and punched up the screen. "The bone-like head, as you can see, comes to a sharp edge like an ax, and it makes a formidable weapon. The two horns growing out of the lower jaw can be used to impale anything behind them with a quick turn of the head. Bones on the knees, elbows, and heels can also be used as cutting weapons. A platelike backbone protects the upper body, and a short tail can deliver a significant blow. Two recessed eyes give tunnel stereoscopic vision that may be more heat-sensitive than light. A set of pointed teeth that have complementary holes in the lower jaw suggests that these animals rip their food apart and swallow it without chewing."

He stopped for a moment, and I took the opportunity to take a sip of my coffee and take a breath. This was a lot to take in, and I needed a moment. Barstow stood for a moment, stirring his coffee with a serious look on his face. I could see he was not anxious to tell me more. I didn't want to press him for more information, but I had to know what concerned him so much.

"What have you to say, Mr. Barstow?" I finally asked.

He stood there for a few moments, put his coffee cup down, and turned to me. He didn't speak immediately, so I just waited for him.

"Captain. These creatures were looking to move into our system. They are intelligent and tough, and they are looking to move into our system."

"But they are dead, Mr. Barstow."

"They were in a battle recently and lost. That gives me cause for concern."

"Do you think that they lost to others of their own kind?" I asked.

"That is most likely to be the case, Captain."

"You mean there is another ship floating around out there."

"Yes, sir. The winner."

The ship's communication light came on, and I opened the

channel.

"We have a probe scan, Captain, and I think you will want to see this."

"We'll be right there, Mr. Anderson."

I could see Mr. Barstow was making last-minute entries in the computer.

"Let's go to the bridge, Mr. Barstow."

"Yes, Captain."

There was much excitement when we arrived, and I asked Mr. Anderson what he had discovered. He set up the playback and warned us to sit down before he started. Mr. Barstow and I sat, and as soon as we were ready, Mr. Anderson started the replay.

"Here we are approaching the planet. Just ahead, the probe takes a hard starboard turn. Look closely. The probe is going to fly right through a bunch of them."

Like a flock of geese, the creatures filled the sky. To see them like this was quite a bit different from our first encounter. It was a beautiful sight. They were really graceful and seemed to be thriving.

"Well, I guess you were right, Mr. Barstow."

"That's what I'm afraid of," he replied.

"Very good, Mr. Anderson. We will leave you at the helm."

I motioned for Barstow to join me back in the conference room. I entered, and he followed behind me.

"Mr. Barstow, just how serious is all this?"

"Serious enough to make some recommendations."

"Such as, Mr. Barstow.?"

"I recommend we blow up the derelict alien ship."

"I'm surprised you would say that. It's our first contact with intelligent life-form."

"Captain. These creatures are a ruthless conquering species, and they are looking for a planet to take over. Earth would be ideally suited to their needs."

"But why blow up a dead ship?"

"We discovered life on that ship. We did not search the whole ship. There may be survivors, and they may have escape shuttles."

"Then you feel that this is necessary?"

"It is imperative that we destroy the ship as soon as possible."

"Then that is what we will do."

He seemed relieved that I complied with his wishes. His argument was undeniably compelling, and I had no other choice.

"Mr. Barstow. I will leave it to you to track down the derelict ship while I go talk to the Rodgers twins to see if we have the ingredients for a bomb."

"I'll get right on it, Captain."

My mind was numb from overload, and although I had reservations about what we were about to do, my trust in my science officer was absolute. When I entered lab 1, the Rodgers twins were busy working.

"Ms. Rodgers," I said, and they both turned around. "I have need of your services."

"Yes, sir. What can we do for you?" said Lorin…I think.

"I need a bomb of sufficient size to blow up a ship the size of… well…the Mobius."

"You want to blow up the Mobius?" they said in unison with some alarm.

"No, I don't want to blow up our ship. I want to blow up the alien ship."

"But that's just a piece of junk," said Lanie…I think. "Why would you want to blow that up?"

"It was recommended by Science Officer Barstow. He said it might be fun."

They looked at each other and smiled like two kids planning a mischievous prank. I didn't know what to make of that, but I smiled back. I started to leave, but they urged me to stay for as Lanie put it (I think it was Lanie), "It'll be ready in a jiffy." I sat down, and in a matter of minutes, they brought it to me.

CHAPTER 37

I returned to the bridge to find Mr. Barstow maneuvering us into range of the wrecked ship. I saw the ship from a different angle this time, and it seemed even larger than before. "I spoke to the Rodgers twins, Mr. Barstow.," I said. "And they prepared a chemical bomb that, they say, will blow the hell out of it."

"They said that?" asked Mr. Barstow with a surprised look on his face.

"Yes, I was shocked."

"How big is the bomb?"

"It's a small package. I think it will fit aboard one of our probes."

I showed him the bomb, which consisted of two glass tubes filled with liquid and were held together with pipe fittings.

"It's a binary bomb."

"A what?"

"Two liquids that, when combined, form a highly explosive gel. The gel is extremely unstable and explodes with the slightest disturbance. Only nuclear explosions exceed the power of a good state-of-the-art binary bomb."

"You never cease to amaze me, Mr. Barstow. How do you know so much about bombs?"

"A misspent youth."

"You're kidding."

"Yes, sir."

We went down to engineering to install the bomb in a probe. I was still curious about how much Mr. Barstow knew about bombs, but I respected his privacy and kept my mouth shut.

The probe module was in a room by itself. The unit required extensive programming to install any additional payload. Mr.

Barstow made light work of this procedure and opened the access panel. A mounting bar was released from the probe, and he fastened the bomb to the bar. He then slid one end of the bar in place and snapped the other end to its alternate-programming receptacle. Then he connected the operating circuit to the probe computer and replaced the cover panel. After pressing the load button, the probe slid into place, and it was ready for launch.

"Very good, Mr. Barstow," I said. "Shall we return to the bridge and blow the hell out of it?"

"By all means, Captain. Let's do it."

When we got to the bridge, the wreck was in clear view. Mr. Anderson was moving us into position.

"That's close enough, Mr. Anderson," said Barstow as he took a seat at the main console.

He entered the launch codes and turned to me. "Ready for launch, Captain?"

"Mr. Barstow. Engage."

He pushed the launch button, and the probe appeared on screen in a trajectory aimed at the wreck. As the probe approached, he switched on the camera and guided it through the large hole in the side of the ship. When we found the right spot, he propelled the probe to lodge in a crack.

"Now let's get some distance," he said.

"We need to be further from the wreck than we are?" asked Anderson.

"This is a powerful chemical bomb," replied Barstow. "We need a bit more distance."

"Fire retros, Mr. Anderson," I ordered.

In a few seconds, we were quite a distance from the wreck.

"Aren't we far enough now?" asked Anderson again.

"Yes, we are," said Barstow. "Now watch this."

He pressed the button that mixes the two chemicals in the binary bomb, and then after about fifteen seconds, he pressed the detonator button. A tremendous explosion with blinding light appeared before us.

"Wow! What was that?" said Anderson.

"A binary bomb," Barstow responded. "We should have been a bit further back."

"I've never seen anything like that," Anderson continued. "Chemical bombs pack quite a punch," replied Barstow.

"That was quite impressive, Mr. Barstow," I said. "And now that we have taken care of that, I need to speak to you in the conference room."

Barstow offered coffee as I sat down at the conference table, but I requested tea for a change. "You asked where I acquired my knowledge of bombs. Maybe you should ask the Rodgers twins."

"I'll take that under advisement, Mr. Barstow."

"What did you want to talk to me about, Captain?"

"I wanted to talk to you about what we just did."

"You mean blowing up the wreck?"

"I mean blowing up a species that we know practically nothing about."

"Sir. From what we have learned, this species is a threat to this system. Everything about them indicates, as I said, a ruthless intelligent conqueror that is looking for a home. Our home."

"I understand, Mr. Barstow, but I…"

"I'm sorry to interrupt, Captain," said Ms. Kane as she stood in the doorway.

"Come in, Ms. Kane. What I was going to speak to Mr. Barstow about concerns you also. What did you want to speak to me about?"

"It's the cell analysis, sir. Of the creature."

"Yes, Ms. Kane, what have you discovered?"

"Well, sir, the cell development is unlike any I have ever seen. I have some photos I can punch up on the main screen." She stepped to the console and punched in the display code. "As you can see," she continued, "the cell structure is more like that of plants. The condition under which this species evolved were extremely harsh. It's a wonder anything evolved. Humans have their skeletal structure on the inside, while insects have theirs on the outside. This creature has a combination of both, and its bones have evolved as weapons. It's ideally suited to be a warrior and can be a vicious adversary. It's omnivorous and would be considered an isomorph."

We sat there for a moment in silence, and then Mr. Barstow looked up. "Have any lingering doubts, Captain?"

"Not anymore."

"Captain?"

"Yes, Ms. Kane?"

"There is something else I need to speak to you about."

"Can it wait for a while, Ms. Kane?"

"I suppose so."

"Then I will see you in a little while."

I turned to Barstow as he was getting up from his chair. He took a last sip of coffee and put it down. "Mr. Barstow," I said

"Yes, Captain?"

"We're going on a short trip."

CHAPTER 38

We were headed to the docking bay, and Barstow hadn't asked where we were going.

"Aren't you interested in where we're going?" I asked. "I know where we're going," he said with confidence. "You do?"

"Yes, we're going to Neptune with a cargo of our pancake-shaped guests."

"How did you know that?"

"It's the right thing to do."

"And I only do the right thing?"

"Yes, Captain."

"Why do you think that is?"

"You always take the advice of your science officer."

"That's because I have the best science officer in the fleet. Don't you agree?"

"Absolutely."

We entered the shuttle through the isolation tube and began the power-up procedure. I programmed the manipulators and opened the sample bin. I checked shuttle communication and contacted the bridge. "Yes, Captain?"

"Is that you, Mr. Blake?"

"Yes, sir. What can I do for you?"

"You can give me a count of the flapjacks we have down here."

"That would be thirty-one, sir."

"Thank you, Mr. Blake."

One by one, I picked up the creatures with the manipulators and placed them in the sample bin. They were sticky on the underside, and a few seemed harder to remove than the others. In about ten minutes though, the job was done.

"Are we ready for launch, Mr. Barstow?"

"Yes, sir."

"Then, Mr. Blake, open the bay doors."

The doors opened slowly, and we eased our way out of the docking bay. And then we were off to Neptune. I locked in our course and put us into overdrive. "What is our ETA, Mr. Barstow?"

"Twenty-two minutes, Captain. Because the atmosphere is so dense, we must enter it slowly, or our flapjacks, as you call them, would fry in our storage bin."

"Very good, Mr. Barstow. Now sit back and enjoy the ride."

I loved flying the shuttle, and this trip promised to be quite fascinating. In just a few minutes, Barstow reminded me to slow my approach. I changed our pitch to allow our underside to take most of the friction of the atmosphere. As we descended further, it became more like surfing, and then we were soaring like eagles. We had to glide around for a few minutes to allow our sample bin to get very cold. When it did, we could hear the creatures stirring.

"You can dump our cargo, Captain."

"Very well. I'll dump them just ahead."

The bin opened slowly, and the creatures hopped out furiously. I circled to see them fly off, but they were nowhere to be found. "They're following us, Captain."

"Well. What do you know about that?"

"These creatures are probably born in flight."

"They think the ship is their mother?"

"They and some of their friends that have joined the pursuit."

"What?"

"Check out your rearview screen."

I looked at the screen and saw hundreds of creatures following us. It was incredible and beautiful. I watched for a moment and then decided to give them a workout. I started executing a turn, and they followed in formation. Next was a total 360, and again they were in

156

perfect synchronization. Even a figure eight did not fool them.

"I think they are enjoying this, Captain."

"So am I, Mr. Barstow."

"There are about a thousand now, Captain."

I increased speed, and again they followed the maneuvers perfectly. The creatures were magnificent in their element of flight. For the next few minutes, I continued our gyrations and then climbed beyond their range. As we looked back, we could see them continue to maneuver in their own patterns, and it was nothing short of spectacular. The cold blue surface of Neptune provided a striking background for the artful display.

"How about that, Mr. Barstow?"

"Quite fascinating."

As I pulled out of the atmosphere, Barstow said something quite surprising.

"Put the pedal to the metal, Captain."

"Mr. Barstow, I can't believe you said that."

"Just trying to fit in, sir."

I cut our speed as we approached the docking doors. Barstow entered the docking codes, and our magnetic guidance system took over. As the lockdown procedure began, we prepared to climb through the isolation tube. Mr. Blake helped us out of the tube and then returned to the docking console. "Mr. Blake, I want the docking bay decontaminated immediately."

"Right away, sir."

Barstow and I returned to the bridge where Ms. Kane was waiting for me.

"Ms. Kane, I will see you now."

We entered the conference room, and I made some coffee. I could see that she was anxious to speak, so I sat down and gave her my full attention.

"What have you to tell me, Ms. Kane?"

CHAPTER 39

She seemed nervous as she took a sip of coffee. "The research I have done with Mr. Greyson has revealed some disturbing facts." She took another sip of coffee, paused for a moment, and then said, "In sick bay, we have had a number of visual problems to deal with, as you know."

"Yes, Ms. Kane. It was your eye drops that cleared up my visual problem."

"Yes, sir, but we continue to have this problem with other crew members."

"And have you given them the eye drops?"

"Well, yes."

"Have they cleared up the problem?"

"Yes, sir."

"Then what's wrong, Ms. Kane?"

"A mistake on my part."

"What do you mean?"

"I thought that the soothing elements of the drops were just clearing up a simple problem of eyestrain. I was wrong."

"Why do you feel that you were wrong?"

"I developed the condition myself. It did not feel like eyestrain. I didn't have a headache, and a few minutes with my eyes closed didn't relieve it."

"So what did you do?"

"The visual distortion was strange, so I did a visual response test with our new VRT-9 unit. The results were very interesting."

"And what was that, Ms. Kane?"

"The machine tests both clarity of vision and wavelength sensitivity. This latter category indicated that my vision exceeded the normal spectrographic range of humans and a clarity of twenty-seven, a reading I have never heard of. With a vision like that, you would see every speck of dust in the air. Distant vision would be difficult through a sea of minute visible particles. My eye drops inadvertently filtered the added range of light, and in about two hours, my vision returned to its normal twenty-twenty."

"Could this be a case of instrument malfunction, Ms. Kane?"

"The tests are repeated automatically to avoid errors due to instrument inaccuracies."

She paused, nervously took another sip of coffee, and pressed a button on the display console. "Captain, it says in your file that at age ten, you broke your left leg when you fell out of a tree. Yet these scans taken just after your visual problems show no sign that your leg was ever broken. And Ms. Grason had a scar on her right shoulder that has miraculously disappeared. All these things may be positive on the surface. I think they are evidence of some kind of virus."

"A beneficial virus, Ms. Kane?"

"Only superficial benefit that indicates their presence."

"Have you spoken to Ms. Dennison about this?"

"No, sir. I've been too occupied with Mr. Barstow."

"Where do you think this virus came from, Ms. Kane?"

"I can't say, sir. We have had a few hull breaches though."

"Those were triggered falsely. And besides, do you know of any virus that can live in the cold environment of space?"

"No, sir."

"Ms. Kane, I must admit that what you have told me is quite intriguing, and I want you to continue your research. If we all have a virus, it has exhibited no lasting adverse effect, and there is no clear- cut course of action. Your arguments definitely indicate something is going on, so I'm depending on you to stay on top of this for there are obviously major ramifications. And since I am first to exhibit symptoms, I will inform you of any further problems. Now go get some sleep."

"Yes, sir. I don't mean to be an alarmist."

"Nonsense, Ms. Kane. Your valuable work is appreciated, and so is your persistence. And I urge you to continue your work, but after some sleep."

"Thank you, Captain," she said as she seemed somewhat relieved.

I was myself in need of sleep, so I took my own advice and retired to my own quarters. Still a little wired though, I again decided to read myself to sleep, but this time, I avoided detective novels.

My next shift of duty started out pleasantly when I met Ms. Kane in the corridor on her way to lab 2.

"I feel much better, Captain. I guess I needed the sleep."

"I told you. Now you go to your station and have a productive shift."

"Yes, sir," she said cheerfully and continued down the corridor. I felt very self-satisfied as I sat alone in the conference room, savoring my first coffee of the day.

"You requested my presence, Captain?" said Barstow as he entered the room.

"Yes, Mr. Barstow. I spoke to Ms. Kane, and she informed me of some interesting anomalies that she discovered."

"Do you mean the plant research we did together?"

"No. The research she did after my visual problem."

"I can't say I know much about that."

"Well, she thinks that the visual problems are the result of a virus."

"Not very likely."

"She also thought it unlikely, but for the lack of a better alternative, she settled for a virus."

"We can isolate a virus with the new equipment."

"Does she know we are capable of this?"

"I'm not sure. We do have facilities that are unique even among advanced research vessels."

"She's been very busy. Her research, her work in sick bay, and her analysis of our visitor. She's had a full plate and may not have had time to familiarize herself with the capabilities of our equipment."

"I will remedy that as soon as possible," he said with a sense of purpose in his voice.

"That would be much appreciated, Mr. Barstow."

"I will get to this right away."

"I'm sure you will, Mr. Barstow."

He walked out the door, and I finished my coffee before I took my place at the helm. I sat down beside Mr. Anderson.

"Captain," he acknowledged.

"How are things on the bridge, Mr. Anderson?"

CHAPTER 40

Mr. Anderson was making a course correction and then turned to me. "Captain, we haven't seen much of you on the bridge lately."

"Well, you know they say that a captain's work is never done."

"I thought that was a woman's work."

"Not on this ship, Mr. Anderson."

Our big screen was aglow with light, and navigation was becoming a problem. More filtering would obscure other objects, so I recommended a decrease in speed.

"Why don't you ask Mr. Brock about his light discriminator?" suggested Mr. Anderson.

"And what is that?"

"I'm not sure, sir, but it might help us navigate."

"Mr. Anderson, get Mr. Brock up here right away. And for that suggestion, I am going to make you a cup of my special blend of coffee."

"Thank you, sir. I feel honored."

I made the coffee and gave Mr. Anderson a cup. "I'll be in the conference room, so tell Mr. Brock to join me there."

I sat down and waited for Mr. Brock while sipping my special raspberry blend of coffee. Mr. Brock appeared at the door, and I asked him to take a seat. I placed a cup of my special blend in front of him.

"You wanted to speak to me, sir?"

"Yes, Mr. Brock. Let me first say that your golf clubs worked great. You're quite an inventor. Mr. Anderson tells me that you have a light discriminator."

"Yes, sir."

"Could you explain how it works?"

"Yes, sir. The device detects microdifferences in polarized

light. It amplifies these differences and de-emphasizes the effects of cross- corona interference. In other words, it dims bright objects without affecting objects of lesser brightness. I like to call it the corona killer. We see light coming directly from objects. We also see light from objects that are reflected from other objects. The off-axis light is what makes objects seem blurry and out of focus. My discriminator only recognizes direct light and rejects all off-axis light, revealing details that were obscured by off-axis glare. This is an oversimplification of a complex process, but it captures the essence of the thing."

"Do you have one of these devices on board?"

"Yes, sir, I do."

"How long would it take to install the system?"

"About thirty minutes."

"Then get to it, Mr. Brock. Oh, you may finish your coffee."

"Thank you, sir. It's excellent."

"I will be on the bridge to test the system, so join me there when you're finished."

I joined Mr. Anderson at the helm just as he was going off duty. Mr. Blake had just reported for duty and was preparing to take over. "May I stick around to see the new system, sir?" said Mr. Anderson. "Actually, I was about to suggest you do just that," I replied.

"And bring Mr. Blake a cup of the special blend."

We all sat for a while, waiting for word from Mr. Brock. "Exactly what kind of system are we getting?" asked Mr. Blake.

"Mr. Brock calls it the corona killer. It differentiates between corona light and direct light and removes the glare from corona effects."

In the next few minutes, we sipped coffee and paced without a word said. It was like a maternity ward with nervous fathers waiting to know if it's a boy or girl. Then my personal communicator buzzed. "Captain. It's Brock."

"Yes, Mr. Brock," I answered.

"Your screen will go blank for just a few seconds and then come back on. As soon as I connect the interface, I will head on up."

The screen soon went blank and came back on again, and we resumed our pacing. I finished the last sip of my coffee when Mr. Brock walked in. "Mr. Brock, we're anxious to see the system in operation."

"Yes, sir. Then let's get to the control console and program it in."

Mr. Blake and Mr. Anderson took their seats at the helm, and Mr. Brock started instructing them on the programming procedures.

"Mr. Anderson. First add the unit to the list of systems."

"Very well, Mr. Brock. What is the name of the system?"

"Corona control system. CCS."

"Entered."

"Make sure that all backup systems are programmed in, Mr. Blake."

"Backups programmed."

"Now since this device is continually variable from 0 to 100 percent, select an attenuator from control bank A."

Mr. Blake looked at Mr. Anderson, and they both agreed on control 9.

"Entered," said Mr. Blake

"Now give the system a few seconds to engage automatic lockdown."

We all waited, looking at the screen, anxious to see how the system operated.

"We're ready, Captain," said Mr. Brock. "Captain, I think you should give the order."

"Mr. Anderson. Engage," I said.

The screen flickered a few times and then stabilized. What appeared on the screen was nothing short of spectacular. It was a tremendous billowy cloud, more beautiful than anything I ever saw on a summer's day in Kansas.

"Mr. Anderson. Go get Mr. Barstow and tell him to get to the bridge, and then go get some rest. You're off duty."

"Right away, sir," he replied and left the bridge.

I turned my attention to Mr. Brock. "I'm going to recommend that your device become standard equipment on every starship in the fleet, Mr. Brock."

"Thank you, Captain."

Mr. Barstow walked in and was instantly drawn to the screen. "You got here very soon," I said.

"I was on my way to the bridge when I saw Anderson."

"Why were you coming here?"

"To discuss a theory."

"What theory?"

"The theory that the source of A-109 is also the source of light. And now we can see it is."

CHAPTER 41

I could see that Mr. Barstow was impressed with the magnitude of the cloud.

"I don't see any plasma reaction," he said.

"That stopped a couple of days ago," said Mr. Blake. "Your duties have kept you off the bridge, Mr. B.," I said. "It's certainly an unusual sight," said Mr. B. appreciatively.

"Would you excuse me, Captain?" asked Mr. Brock. "I must get back to an experiment that I'm working on."

"Yes, Mr. Brock. And keep me informed of anything you are working on."

He left the bridge, and I turned to Barstow as he stared at the screen. "It just doesn't look like it belongs there."

"It doesn't," Barstow replied. "Cloud formations are the results of air temperature and pressure in a gravitational ecosystem. No forces in space could account for this formation."

"What are you saying, Mr. Barstow?"

"Something is controlling it."

"Why would that be? Or more to the point, who would that be?"

"Maybe the entity that did so much damage to the alien vessel manufactured the cloud."

"For what purpose?"

"Camouflage?"

"Mr. Barstow, I want someone watching that cloud at all times until further notice. You will abandon all but your bridge duties for the time being. Now I am going down to Ms. Kane to tell her why you will not be joining her in her research. And get another pair of eyes on that screen. Mr. Decker, if he's busy, tell him to drop everything and get up here on the double."

I stepped lively as I made my way to lab 2. Ms. Kane was at her station as I entered. "Ms. Kane. Mr. Barstow will not be joining you just now. I have assigned him other duties."

"But, sir, he's been a great help."

"I understand, Ms. Kane. However, something has come up of dire importance, and I need him on the bridge."

"What is the emergency?"

"I don't have time to go into that now. You'll just have to continue without him for a while."

I started to leave when Ms. Dennison approached me. "Captain, I must speak to you about something."

"I really don't have the time now, Ms. Dennison."

"Then I suggest you make the time, Captain."

"Very well, Ms. Dennison. Get to it."

"It's Mr. Basser."

"I'm sorry, I've been so busy. How is he doing?"

"Strangely enough, very well."

"What do you mean strangely?"

"He's secreting a green substance from his leg."

"You mean he has gangrene?"

"No, sir. In fact, the substance seems to be of benefit."

"What do you mean of benefit?"

"This is going to sound crazy."

"Well, spit it out."

"I think Mr. Basser is growing a new leg."

"Ms. Dennison. I am not a medical officer, but I know that nobody grows back severed limbs. Now what is going on with this green substance?"

"Cell reparation."

"Are you kidding? I want to see Mr. Basser."

She led me to cubicle number 3, where Mr. Basser was fast

asleep. "Is he on sedative?" I whispered.

"No, we don't give him anything. He seems to sleep naturally when he needs to."

"Let's see the leg."

"I warn you, it's not a pretty sight."

She pulled back the sheet and revealed just what she said: a disgusting sight.

"Oh my god!"

"As you can see, Captain, his leg is at least four inches longer than a week ago." She covered the leg, and we went back to the lab. We sat at her research console, and we observed the result of her microscopic examination. Computer codes were entered, and the display screen lit up with a sample of the green substance. "These are some of the green cells taken from Mr. Basser's leg. I took some cells from his arm and added them to the green substance. After a time, the green cells started changing. They took on the characteristics of the human cells. So accurate is this transformation that our analyzer can't differentiate."

"And you're worried about this?" I asked. "Be happy for Mr. Basser while you study the hell out of this. I'm getting tired of saying this, but keep me informed of your progress. And cheer up, for Christ's sake."

"What other anomalies were we in for?" I asked myself out loud as I walked the corridor. I need to work off some tension, so I stopped by my quarters to change into my sweats and headed to the handball court. Ms. Yates was there, slapping the ball around.

She stopped when she saw me and walked over. "Captain, I haven't seen you here in a while."

"No, Ms. Yates. I've been a tad busy. The polo season, you know."

"Are you putting me on, Captain?" she asked coquettishly.

"Now would I do a thing like that?" I replied.

"In a minute," she retorted. "Care for a game, Ms. Yates?"

"Are you challenging me, Captain?"

"That's right."

"You're on, Captain."

I slapped the ball hard and picked up the first point. She looked surprised but started playing with more intensity. I scored another point and then another and still another. She finally scored a point, and I came right back and scored again. At the end of the game, she seemed exhausted.

"Now that's a real game," she said, smiling. "How do I know you didn't throw the game?"

"Would I do that, Captain?"

"In a minute."

CHAPTER 42

I fell into bed, exhausted, in every sense of the word. The avalanche of amazing discoveries was pushing my mind to the point of overload. Being surrounded by some of the most talented scientists was certainly keeping me on my toes, and though it was intellectually stimulating, it was also gruelingly demanding. The challenges that lay ahead of us were unprecedented. I was confident that my handpicked crew would deal with whatever we encountered. It was my contribution that worried me.

Life goes on though, and the activities of everyday life, even aboard a starship, were a welcome distraction. I was getting a craving for another of the fine meals prepared by the Rodgers twins, so I visited the chemistry lab. I also wanted to congratulate them on their bomb construction. They were, as I expected, busy working when I walked in.

"Captain," they said in unison when they saw me from across the room.

"Ms. Rodgers. I wanted to thank you for putting together the binary bomb. It functioned magnificently. I am curious to know how you came by this knowledge."

"Our father is Mr. Rodgers," said Lorin, I think.

"Yes?" I said with what must have been a dumb look on my face.

"Of Rodgers Demolition," said Lanie, I think. "Oh. That Rodgers. Now I understand."

"It's how we developed an interest in chemistry," they said again in unison.

"I was wondering if you could put together another of those wonderful meals this evening?"

"It would be our pleasure, Captain."

"Then I'll see you at dinner," I said happily.

I turned to leave only to see Ms. Dennison, who was sitting at her station. She motioned for me to come over and join her, so I did. "Ms. Dennison, is there something you need to speak to me about?"

"Yes, sir. I ran another test to see if the green substance combined with cells that were not Mr. Basser's."

"And what were the results?"

"I took a cell sample from myself and added them to the green substance."

She took a deep breath and for a moment, seemed to lose her train of thought.

"Ms. Dennison?" I said.

"Oh. Yes," she said as she regained her composure. "When I added my cells to the green substance, it consumed them."

"And what conclusion did you draw from this?"

"The substance is being produced by Mr. Basser. His body, that is."

"How could that be, Ms. Dennison?"

"I don't know, Captain. This is totally baffling," she said in frustration.

"Ms. Dennison. I want you to go to your quarters and take a nap until dinner, when we will once again enjoy a feast prepared by the Rodgers twins."

"But, Captain—"

"That's an order Ms. Dennison. You've got to relax. Now come with me."

I grabbed her hand and walked her down the corridor to her quarters. "You're working yourself into a tizzy. Now I'll be back in a couple of hours to escort you to dinner. We'll have a bottle of wine, enjoy the meal. Maybe we'll even have a laugh, and then I will bring you back to your quarters for a good night's sleep."

"All right, Captain."

"Cheer up. We'll have a good time."

"Perhaps you're right."

"Of course, I'm right. I'm the captain."

I left her at her door and headed back to my quarters and took my own advice by catching a couple of hours of sack time. I slept well, and upon awakening, I prepared for my dinner date. My best uniform was in slot 6 in my closet, and I hoped it would impress Ms. Dennison. It had been a long time since I had been on a date. Ms. Dennison was quite lovely, and her wit made her an interesting dinner companion. Like some teenager, I found myself nervous and fumbling. After a shower and shave, I splashed on some of my favorite cologne. Spit-shined shoes completed the ensemble. Someone was at my door, so I opened it to reveal a stunning Ms. Dennison. She had on a beautiful black evening gown with matching shoes.

"I was supposed to pick you up, Ms. Dennison."

"I guess I was too anxious."

"Well, don't you look stunning?" I said with complete sincerity.

"You're not so bad yourself, Captain."

I extended my arm, and we strolled down the corridor to the ship's galley.

"And how was your nap?" I asked. "Remarkably refreshing."

"I know this delightful bistro that is just down the street."

"Captain. I'm sure you know all the best places around here."

"What are you in the mood for?"

"Food."

"I think they have that."

Everyone noticed as we entered the galley. Mr. Clark acted as our waiter and showed us to our table. "Care for some wine, Ms. Dennison?"

"Certainly, Captain."

"Bring us a bottle of your best, Mr. Clark."

"Yes, sir," he said and left.

"What do you think of this little place?" I asked. "It's charming, Captain," she continued being coy. "I only bring special people here."

"Like Ms. Yates?" she said in a jealous tone. "What do you mean?" I quickly replied. "Everyone knows she has a major crush on you."

"That's ridiculous," I said, surprised. "She's always chasing you around."

"That's not true."

"Well, she's always on the handball court waiting for you."

"How do you know that?"

"It's a small town."

"It's not true."

"When you played handball last, who did you play with?"

"Ms. Yates."

"See what I mean?"

CHAPTER 43

Mr. Clark returned with the wine and poured us both a glass. "I would like to propose a toast."

"And what are you toasting, Captain?"

"To my lovely dinner companion."

"Captain, I didn't know you were so romantic."

"I've always been romantic."

"I thought there was a rule about fraternizing between crew members."

"Is that what we're doing, Ms. Dennison? I thought we were having dinner."

"Of course, Captain."

She took a sip of wine and sat back as if to relinquish the conversation. I sat back as well to see how she would react. A few moments passed as she seemed to be speechless.

"Why are you doing this, Captain?"

"Doing what?"

"Not speaking."

"Most people would consider that a good thing."

"Well, I don't. I want to be entertained."

I looked up to see that Mr. Clark was bringing our food. "Why don't we eat first?" I suggested.

"Very well, Captain. But afterward, you owe me some first-class entertainment."

"And you shall have it, Ms. Dennison. First though, let's sample the culinary offering."

Our dinner was superb, and it helped improve the tone of our conversation. Even an occasional smile found its way to her face as she slowly fell under the spell of the candlelight atmosphere.

"The food is great," she said.

"I am more than happy to have arranged it," I answered. "You certainly know how to treat a lady, Captain."

"It's my pleasure, Ms. Dennison."

"You're really quite charming, Captain."

"Are you making a pass at me, Ms. Dennison?"

"Ms. Yates isn't the only female with hormones aboard the ship."

"And I thought you were strictly a career-minded medical officer."

"All work and no play makes Lisa a dull MO."

"I'll keep that in mind."

When we had finished, we had one final glass of wine and left the galley. She noticed that we were not walking in the right direction down the corridor. "Captain. My room is the other way."

"Yes, but the entertainment planned for the evening is in this direction."

"Where are we going?"

"You'll find out momentarily."

We came to the docking bay doors, and I pushed the access button. The doors opened, and I took her by the hand and escorted her to the shuttle. She had never been aboard a starship shuttle before. "I was not aware that the inside of a shuttle was so tastefully decorated. The gray-and-black panels are very attractive."

"The shuttle is designed to be very comfortable."

I logged in with Mr. Matson. He answered, and I asked him to open the bay doors on my signal. "You're going someplace?"

"Yes. We're going to Neptune."

After entering the operation codes, I engaged the boosters, and we were on our way.

"Oh my. This is great, Captain."

"I thought you might like it."

I made a few more entries, and she watched with curiosity.

"What did you just do?" she asked.

"I engaged the differential navigation system."

"What is that?"

"It's a system that prevents overreaction by a pilot."

"What do you mean?"

"If I were to make an irresponsible navigational decision, the system would override with proper course changes to prevent catastrophe."

"In other words, anybody could fly this crate."

"I don't think I would have put it quite that way, but you could say that."

We approached the planet and started cutting through the atmosphere. The dense cloud cover limited our view, and then we broke through. And there before us was a formation of flying creatures.

"Oh my. They are beautiful," she said. "Yes, they certainly are," I agreed.

Hundreds of creatures of all sizes were flying about in perfect synchronization. The sight was out of a fantasy, and we had front row seats. Ms. Dennison's eyes were wide open as she took in the incredible scene.

"Oh my. This is spectacular, Captain."

"Yes, it certainly is, Ms. Dennison."

After about thirty minutes, I pulled away and headed home.

She sat back in silence, stunned at what she had just seen.

"Will this suffice for an evening's entertainment, Ms. Dennison?"

"Oh yes, Captain. I can't think of anything better. It was magnificent."

She continued the accolades as we approached the docking bay. I made the proper entries to engage the series of automatic docking and lockdown maneuvers. I opened the shuttle door and helper her out then escorted her back to her quarters.

"This has been a marvelous evening, Captain," she said sincerely. "My pleasure, Ms. Dennison."

She kissed me on the cheek quickly and entered her quarters. I stood there for a moment and then walked away. Mr. Clark was coming down the corridor, and I stopped him.

"I want to thank you for acting as our waiter for the evening, Mr. Clark."

"You obviously needed a waiter, so I was happy to do it, sir. Oh yes, I meant to speak to you about a staff meeting that was canceled because of all the excitement that has been going on here lately. Should I reschedule?"

"No, Mr. Clark. I will inform you of the next staff meeting."

"Very well, Captain. I hope you had a nice evening."

"I had a fine evening, Mr. Clark. Thank you."

When I retired to my quarters, I thought about the events of the evening. Ms. Dennison seemed back to her old self, and I was happy to have provided a distraction from the stress of her expanding duties. I decided to get a jump on the next day's schedule and entered a few notes to the captain's log. Because of all the excitement, I was behind in my entries, and since I was wide awake, I thought it would be a good time to catch up.

CHAPTER 44

On my way to the bridge, I still had the events of the night before on my mind. Mr. Anderson was on duty, and I took my place at the helm.

"Anything to report, Mr. Anderson?" I asked. "No, sir. All is in order."

"The cloud. Any activity?"

"Well, sir. It is undulating very slowly, as if like a normal cloud."

"Which should not be. With no forces acting upon, it should just dissipate."

"That's right, Captain," said Mr. Barstow as he sat down beside me. "This cloud has maintained core density for as long as we have been monitoring it. The question is, why is there a core density in the first place?"

"Can you explain the cloud, Mr. Barstow?"

"The cloud does not disperse as any dust cloud would in the vacuum of space, and we are not sure what holds the cloud together. There is no force that is being applied that we can detect, and it has no magnetic properties."

"Is there any connection between the cloud and our alien visitor?"

"No, Captain."

"Do we know if the plasma field had any effect on the cloud?" asked Mr. Anderson.

"It may have made more."

"You think that plasma makes the powder?" I asked. "It's just a theory."

"What do you think comes in contact with plasma that turns into an amorphous powder?"

"I have no idea. However, since plasma is in such plentiful supply and the cloud is tremendous, a quantitative relationship is obvious. This cloud did not travel here in its present state, for as stable as it appears, it is actually a delicate balance."

"If I remember correctly, you said you had tagged the powder with a tracer. Has that shown up?" I asked.

"No, sir."

"Well, Mr. Barstow, continue your work and let me know immediately of any new discoveries."

The meeting ended, and I sat for a few minutes, thinking about what was just said. I decided to do some research of my own, so I paid a visit to lab 2, where I found Ms. Cooper.

"How are you on this fine day?" I asked. "Captain. I'm fine. What brings you to lab 2?"

"I need some information. If I may join you for a few minutes."

"Of course. Have a seat. Now how can I help you?"

"Ms. Cooper, when you first collected samples of the powder, you had problems keeping track of the substance."

"That's correct."

"I would like you to call up some information regarding those early experiments."

She started entering data on her control console. "What information do you require?"

"We have an array of sensors that give us the exact internal volume of the chamber. I want you to punch up the displacement volume of the chamber when we left the Stockton." The data was entered, and the information came up on the screen.

"Now, Ms. Cooper, call up the internal volume of the chamber just after you lost track of the sample."

A different figure appeared on screen. "We already know this, Captain," she said.

"That's correct, Ms. Cooper. What I want you to do is give me a reading on the chamber for right now."

"But, Captain, we have the creature in the isolation chamber."

"Then measure the total displacement value of the creature and add it to the remainder." She entered the data, and in a few seconds, the figures appeared on the screen.

"That's it," I said. "They are identical."

"What do you mean, Captain?"

"The internal volume of the isolation chamber has returned to its original size."

"What does that mean, sir?"

"Since volume is constantly monitored, let's find out when the chamber returned to its original size." Again she entered the data and the figures. Fourteen hours, thirty-two minutes and eleven seconds appeared on the screen.

"Ms. Cooper. Thank you very much."

I made a quick exit and made it to the conference room. Mr. Barstow was on deck when I got to the bridge, and I asked him to join me in the conference room. I sat down quickly, and Mr. Barstow, sensing the urgency, did the same.

"Mr. Barstow. Something you said stuck in my mind, and I did some research of my own."

"What did I say, Captain?"

"You said that the substance traveled here from outside the system. That suggested to me a thought process."

"That's very astute, Captain."

"What if the amorphous powder is actually a life-form?"

"You mean like a symbiotic substance that is looking for a host?"

"Well, no. But I like your idea better."

"It would explain a number of things. Like why the cloud stays together and why it glows."

"Why is that, Mr. Barstow?"

"To attract hosts."

"Oh yes, of course."

"That's very good, Captain."

"I came to it from hanging around you, Mr. Barstow."

CHAPTER 45

I took time to make coffee for it felt like this meeting was going to take some time. Mr. Barstow was taking notes and seemed enthusiastic about this revelation. In a few moments, he looked up and asked me a question.

"Is there anything else, Captain?"

"Yes. I checked out the volume of the isolation chamber, and it has returned to its normal size."

"Suggesting that whatever fused with the chamber passed all the way through, causing hull breach alarms to trigger."

"I never thought of that."

"I did. It's my job. And there is something else."

"What's that?"

"If it passed all the way though, it did not use its identity, and it can differentiate between organic and inorganic substances."

"That's amazing, Mr. Barstow."

"It also explains some other occurrences."

"What specifically?"

"When Ms. Kane suspected we had a virus, she was not far from the truth. We had all been contaminated with the airborne substance."

"Is there anything we can do about that?"

"Not likely. If there are any negative effects from the substance, they haven't shown up."

"What about the visual problems?"

"That was a temporary problem that I don't think has any long-term ramifications."

"Why do you feel that way?"

"The evidence."

"What evidence?"

"Ms. Kane brought to my attention that many preexisting conditions of crew members were not showing up in medical scans. A broken leg that you had as a kid has not shown up, scar tissue from an injury on Ms. Grason's arm has disappeared, Mr. Palasco's asthmatic symptoms have vanished, and Mr. Basser seems to be growing a new leg."

"Ms. Dennison tried adding some of her own cells to the green substance from Mr. Basser's leg, and her sample was consumed."

"That's because the substance had come in contact Mr. Basser before Ms. Dennison added her own DNA sample. The substance adapts to the first organic thing it contacts. All other DNA is considered an invasion and will be attacked."

"It sounds as if you have this all worked out, Mr. Barstow."

"I've been working it out for some time. Your discovery just tied up a few loose ends."

"What about the time we were subjected to our deteriorated atmosphere?"

"We had all had contact with the substance at that time, and it enabled us to breathe the available atmosphere."

"So we were all infected?"

"I'm not sure I would use the term infected, but yes, sir. We were."

"What about the magnetic flux, Mr. Barstow?"

"We had experienced several hull breaches before we came in contact with the magnetic flux, so we must have been infected before that. What made you think of the isolation chamber, Captain?"

"Well, if the powder could travel through space and through solids, the chamber should have returned to its normal size."

"I'll bet you know how long the process took."

"Fourteen hours, thirty minutes, and eleven seconds."

"Captain. I believe you are becoming a scientist."

"Your influence is inspiring, Mr. Barstow, but I don't think that I should start preparing for the Nobel Prize just yet."

"You may be right, sir."

"Mr. Barstow, do you think there will be any ill effects of this substance either now or long-term?"

"No, sir. The substance has sought a symbiotic host. It would not be in its best interest to harm its host. In fact, it has improved our health in ways that we have yet to discover. In space, it must maintain a powder form to survive. When it contacts anything, it probes for life, and when it finds life, it combines with its DNA characteristic and repairs, maintains, and defends its host."

I took the last sip of coffee, and so did Mr. Barstow. "If there is no further business, I suggest we get to our respective duties," I said. He left the conference room, and I sat for a few minutes to gather my thoughts after such an intense meeting. I stood up and headed to communications.

Ms. Yates was there by herself, and she smiled as I entered the room. "Captain. Haven't seen you around for a while."

"I've been a bit occupied with mundane chores."

"Yea. I've heard about those chores."

"I wanted to speak to you about the communication project."

"I assure you, Captain, I am better prepared this time."

"If you need any help, I can arrange for Mr. Barstow to help you."

"I'd appreciate that, Captain."

She looked up to see Mr. Clark enter the room.

"Mr. Clark. I would like to schedule a medical scan for every crew member including yourself within the next seventy-two hours."

"Yes, sir. Would it be convenient for you to be first?"

"That would be fine, Ensign. I have business to attend in sick bay in about an hour."

CHAPTER 46

"I'm ready for my scan," I said as I entered sick bay. "Take a number," said Ms. Dennison sarcastically.

"I'm glad to see that you are in top form today, Ms. Dennison."

She made a few computer entries and then turned her attention to me. "Now what seems to be the problem, Captain?"

"I have scheduled a medical scan for all crew members in the next seventy-two hours, and I came down to inform you."

"Gee. Thanks, Captain. I was wondering what I was going to do with all that spare time."

"I know you're busy, but we need to monitor closely the health of all crew members. And that includes the medical staff.

"Is this your idea?"

"No, Ms. Dennison. Mr. Barstow and I have decided to call for the scans due to some new information regarding the effect of A-109."

"Is there is anything I should know, Captain?"

"I suggest you get together with Mr. Barstow and Ms. Kane to get updated."

"I will take care of that after I've run your scan," she said seriously.

"I suppose you want me to get on the scanning table?"

"That's the way it's done."

"I expect you to run one of these on yourself, Ms. Dennison, and submit the results to Mr. Barstow with all the others."

"Lost confidence in my ability, Captain?"

"Of course not. We are just looking for something special."

"Yes, Captain. Now get up on that table. That's an order."

I did as she said, and we began the scanning process. After calling up my previous scan for comparison, she started the procedure. In about four minutes, it was completed.

"You may get up, Captain."

"Thank you, Ms. Dennison. I'll let you get back to your duties."

"Before you go, Captain, there is something I'd like you to see."

We walked to the patients' room area and entered Mr. Basser's room. He was finishing a small cup of pudding.

"You like the hospital food, Mr. Basser?"

"It's not bad, Captain."

"How are you today?"

"I'm fine, sir."

"Don't be so modest, Mr. Basser," said Ms. Dennison. "Get up and show the captain your new leg."

He got up and stood on two fully functional legs. "Mr. Basser," I said, startled at the sight. "You're certainly looking well."

"Quite a doctor we have here, don't you think, Captain?" he said, smiling.

"I think you're right, Mr. Basser."

"I'm ready for duty, Captain."

"What do you think about that, Ms. Dennison?"

"I think some light duties on the bridge might be in order."

"Very well, Mr. Basser. Report to the bridge in one hour."

"Thank you, Captain. I can't wait."

Ms. Dennison and I returned to the lab where we reviewed Mr. Basser's case. "That's absolutely amazing, Ms. Dennison. His leg looks perfectly normal."

"And it tests perfectly normal. He—like you, Captain—had a broken leg as a kid. It was not the one severed, and there is no sign of a break on scan. In every way conceivable, the tests are textbook normal, except for one thing."

"And what is that?" I asked.

"No one is symmetrical. The left side of the body is different than the right side. Mr. Basser's new leg is an exact mirror image of the other, right down to the number of hairs."

"How has he adjusted to the new leg?"

"He walks as if he had never lost a leg."

"Were you able to save any of the green substance?"

"Yes, sir."

"What have you discovered?"

"I believe that when the powder makes contact with DNA, it commits to that organism and will defend it. If some of it is separated from the host, it is reluctant to combine with anything inorganic, so it can be contained. If introduced to a different organic sample, it will connect to the new DNA. If it is not introduced to a new DNA sample within four days, it will dry up, turn back into a powder, and pass through any inorganic container in search of something organic."

"What tests are you running now?"

"I have taken cells from myself and combined them again with the substance that had been separated from Mr. Basser for forty-eight hours. Unlike the first experiment, the substance combined with my cells."

"Ms. Dennison. I want you to get together with Mr. Barstow and inform him of your findings. If there is nothing further, I must get back to my duties."

Engineering was my next stop, and I was anxious to speak to Mr. Brock. I really felt that I should spend more time in engineering. The fast-paced activity, the feeling of power and control, and just getting your hands dirty were quite stimulating. At the helm, even great speed was not perceivable, and hours spent in front of a screen where nothing seemed to change could make helm duty tedious.

When I got to engineering, Mr. Decker and Mr. Langer were working on a unit that would improve the efficiency of our main thrust engines. Mr. Matson had managed to boost our power system by 5 percent. Ms. Grason was continuing to refine and enhance the performance of our robots, and Mr. Brock was working on a project that I was curious to see. Mr. Langer was running tests from his station and looked up to see me approach. "Captain. What can I do for you?"

"You can tell me where Mr. Brock is conducting his research."

"That would be engineering lab 2."

"I walked over to lab 2 where Mr. Brock was hard at work on a strange-looking device. He didn't notice me for a few minutes, and then I spoke. "Mr. Brock, I thought I'd come down to see what you were working on."

CHAPTER 47

He walked around to the other side of the table before he started his explanation. He made a few final adjustments and an entry at his computer station. "Captain. Spectral analysis has been a useful procedure that, for a long time, has given us some special abilities that have expanded our technology. White light is incoherent and is made up of interference patterns. Both direct light and corona light can provide important information. Direct light gives us what we perceive as focus or precise image recognition."

He paused to take a sip of coffee and then resumed his enthusiastic explanation. "Corona or interference pattern light that we perceive as glare or reflection can actually provide us with more information than direct light." He paused again and prepared for what seemed to be the wrap-up. "If we use a high-speed sequencer to alternate between preselected amounts of direct light and corona light, we might see with an accuracy that has never been dreamed of."

"And you are adding this new circuit to our corona killer?"

"No, sir. That circuit is already in the corona killer."

"Then what is the new project you are working on?"

"Now that's really special. That little unit I was working on when you came in will increase the efficiency of our air conditioning system by 10 percent. Now that's something."

"Yes, Mr. Brock, that's really something. You get to that right away."

I left engineering scratching my head and wondering what goes on in the mind of a genius. My next stop was lab 2 for a visit with Ms. Cooper and Ms. Kane. I was anxious to see what they had to report. When I arrived though, there was no one there. I stepped over to the isolation chamber and looked at the creature. It laid there like some kind of horrible statue, frightening yet fascinating. An eerie feeling came over me as I stood there.

"Captain," said Ms. Kane as she entered the room. "Oh. Ms. Kane, you scared me."

"I'm sorry, sir. What brings you to lab 2?"

"I just came to see how our guest is doing."

"We have more information for you."

"About the creature?"

"We don't need to refer to it as creature anymore. It's an isomorph."

"Yes, I believe Mr. Barstow called it that. What does it mean?"

"Solid matter generally is of crystalline structure and displays an orderly atomic arrangement."

"Which means?"

"Solids that exhibit no crystalline structure, like glass, are called amorphous. This is a silicon-based life-form that has developed the ability to secrete a glass-like protective coating when it is injured, and for lack of a better term, we chose isomorph. It has also developed the same temperature and atmospheric requirement of a carbon-based life-form." She paused for a moment when Ms. Cooper entered.

"Captain. It's good to see you in lab 2. We have been busy and have a lot to report."

"I see, Ms. Cooper. Ms. Kane has been telling me."

"We have some interesting pictures to show you."

"What kind of pictures?"

"We have a fiber optic microscan probe connected to the creature."

"Don't you mean isomorph?"

"Oh. She told you the classification."

"Yes, Ms. Cooper."

"Well. Let's get to the examination."

She entered codes at the medical console and then sat back as the computer established a link to the medical probe. The screen then lit up with a magnified view of the cell structure of the isomorph. "As you can see, Captain, the structure is quite unique like nothing we have ever seen."

"I see, Ms. Cooper. Continue."

"There are perforations uniformly scattered all over the external malleable tissue. We believe this is the way it breathes."

"What else have you discovered?"

"I think Ms. Kane should continue. She's the expert on body chemistry."

"Very well. Ms. Kane, would you continue?"

She made a few data entries and adjusted the power of the microscope. "When injured, the isomorph secretes a clear substance that hardens around the wound until it heals."

"Why is he totally encased in the substance?"

"It was hit by enemy fire many times."

"Have you analyzed the clear substance?"

"Yes. It appears to be glass."

"You mean glass like our view ports?"

"Not exactly. There are a couple of other unidentified ingredients."

"Have you conducted a magnetic scan, Ms. Cooper?" I asked.

"Yes, sir. The isomorph has no iron content or any other substance with magnetic properties."

"Are there any conclusions that you can draw from this?"

"Not with our data bank," said Ms. Kane. "There are no practical references."

"Well, you'll have to start writing a new book," I stated. "Yes, Captain. I think you're right."

"What do the Rodgers twins think of this?"

"They are in lab 1 now running every known test," said Ms. Cooper. "They seemed a little frustrated. They don't like to be stumped."

"Maybe I should go up and give them a word of encouragement," I said.

"That might be a good idea," said Ms. Cooper. "I was just up there, and they were arguing."

"Are you serious?" I asked. "Yes, sir."

I noticed a distorted picture on the view screen and called Ms. Kane's attention to it. She tried to clear it up when I looked through the chamber window and saw the isomorph move. "Oh my god!" I said as I watched the giant seven-foot alien stand up.

CHAPTER 48

All three of us stood there, not knowing what to do. Ms. Kane made a quick entry in the medical console. "It's okay," she said. "He can't get out of there."

I tapped my communicator badge. "Mr. Barstow. Get to lab 2 immediately."

The isomorph moved slowly as his glass exterior coating shattered and fell to the floor" It stood still for a moment with its head hung down, as if praying.

Mr. Barstow ran in and saw what was going on. "How long has it been conscious?" he said.

"Just a few seconds," said Ms. Kane.

We all stood there looking at the amazing specimen, wondering what it would do. I looked at Barstow to see what his reaction was. "Its head is rising," he said.

"Stay still, everybody. He can't get out of there," Ms. Kane repeated. Everyone was quiet for the next minute.

"The alien has some kind of weapon, but he can't see us at this angle," said Barstow.

"Let's get to the door and shut down lab 2," I said. We got to the exit and locked down the lab. "I thought that thing was dead."

"There was no life-form reading on our instruments," said Ms. Kane.

"Let's get to the bridge," Mr. Barstow suggested.

We all headed to the bridge with a measure of urgency. "You said you saw a weapon?" I asked Barstow.

"Yes, sir. In his right hand."

"Any idea of what it can do"

"I'm sure it can do a lot of damage."

When we got to the bridge, Mr. Anderson was on duty as we all took seats at the main console. "Mr. Barstow," I asked. "What are your recommendations?"

"We must get him out of the lab to a section with an external hatch."

I turned on the camera in lab 2, and the alien came up on the screen.

"It's hardly moved," said Ms. Kane.

"It's disoriented and is wondering where it is," answered Barstow.

"Shouldn't we strike now?" I asked.

"You read my mind, Captain."

He opened the door to the isolation chamber. The alien did not move. "It senses a trap, and I think it will take time to evaluate its options."

"That gives us some time, Mr. Barstow," I said. "That will give us time to suit up."

"Once again, Captain, you read my thoughts."

"Then let's get to it."

We stepped over two of our supply cabinet of zero-atmosphere suits. I slipped into my suit as Barstow was doing the same. I put on my helmet and locked it into place.

"Mr. Anderson," I said. "Do we have full com-link?"

"Yes, sir, we do."

"Mr. Barstow. Do we have full com-link?"

"Yes, sir."

"Mr. Barstow, is there anything on board that could be used as a weapon?"

"I believe we have some hammers in engineering."

"Let's get down there," I responded.

"Mr. Anderson, let engineering know that we are on our way."

"Yes, sir."

The suits were heavier and more uncomfortable than I remembered. Our urgent pace got us to engineering quickly, and we were met by Mr. Decker, Mr. Matson, and Mr. Brock.

"Captain," said Mr. Brock."

"I've got the perfect hammer for you. It has a small flat head and over seven inches of pick ax in the back."

I looked at the hammer and then Barstow. He looked up and nodded his approval.

"We approve of your choice, Mr. Brock. We'll take two."

He handed each of us a hammer, and we headed to lab 2. "Should we discuss a plan, Mr. B?"

"No. Too many variables and an enemy that is too unpredictable. We must keep our wits about us." Just ahead was the door to lab 2. We proceeded with caution and approached slowly.

"Mr. Anderson," I said over the suit's communicator. "Yes, sir."

"I want you to open access-way doors 11 and 12. Did you hear that, Mr. Barstow?"

"Yes, sir, I heard."

"I want to lure the alien out of the lab through door 3 and into the access way through doors 11 and 12. This will bring us to external hatch 41. Mr. Barstow and I will attach our lines to the tether hooks, and on my signal, I want you to blow the hatch. Did you get that, Mr. Anderson, Mr. Barstow?" They both acknowledged, and we set the plan to action. "Mr. Anderson. Open door 3 of the lab."

He opened the door, but there was nothing to be seen. We stood there for a couple of minutes. Mr. Barstow suggested we go through the two access doors and attach our two tethers before proceeding. I agreed and ordered Mr. Anderson to open access doors 11 and 12.

We went to the outer access way, quickly attached our tether lines, and waited. "Mr. Anderson, where is the alien?" I asked.

"He is still in the lab."

"What's he doing?"

"He's looking around."

"Close the two access doors and let me know if the alien moves." Mr. Barstow and I stood there with our tether lines in one hand and the hammers in the other. We waited for something to happen, and then I heard Mr. Anderson.

"He's coming through the lab door."

"Let me know when the alien is standing in front of the access door."

"He's headed there now. No, he stopped and is looking around. He's moving away from the door. He turned around and is facing the door and is still is looking around."

"Open the door, Mr. Anderson," I said. He opened the door and said nothing for longer than I liked. "What's going on, Mr. Anderson?"

"He's just standing there. No, he's moving to the door, and he just went through."

"Close the door behind him." I waited a few seconds. "Did he react to the door closing behind him?"

"No, sir."

"Does he still have his weapon?"

"Yes, sir, and he's pointing it at the door."

"Open it, Mr. Anderson."

I tightened my grip on the hammer and braced myself for what I was about to experience. Mr. Barstow had also prepared for the moment at hand. Nothing happened for the next two minutes.

"Can you see anything?" I said. "No, sir."

"Captain, he's just standing there," said Anderson.

CHAPTER 49

I was getting very nervous, waiting for something to happen, when Mr. Anderson spoke.

"He's moving forward."

I could see his hand pointing the weapon forward, and then he turned to me and was about to fire when Barstow knocked the weapon out of the alien's hand.

"Blow the hatch!" I yelled.

The hatch swung open, and the air pressure pulled the alien off its feet. It reached out and grabbed Barstow's hammer. Barstow was almost pulled out of the ship. But he let go of the hammer, and the alien was drawn to the hatch. The alien jerked its head back and embedded its ax-like skull into the floor. I quickly moved over to the alien and tried prying the skull loose. I could feel the hand of the alien grab my arm with tremendous force. Barstow kicked the head of the alien, dislodging it from the floor.

We were all pulled out of the ship. I could not break free from the grip of the alien, so Barstow took my hammer and struck its arm. The grip loosened, and the alien was pulled away.

"Captain," I heard Mr. Anderson say. "Is everything all right?"

"Yes, Mr. Anderson. Reel us back in."

We felt the tug of our tether lines as we were pulled back to the ship. "Captain," said Mr. Anderson. "Brock and Decker have suited up and are waiting at the hatch door to help you back on board." In just a minute, Mr. Brock reached out and pulled me in while Decker helped Barstow.

"Close the hatch, Mr. Anderson," said Brock, and the hatch closed as ordered. "Pressurize, Mr. Anderson," he said again with authority.

The room filled with air as we waited for the corridor door to open. He then helped me to sick bay, and I noticed that we lost Mr. Decker and Barstow.

"They're just behind us, Captain."

We entered sick bay, and I was helped up on a table. Brock helped me take off my helmet.

"Mr. Brock," I said. "Your hammers worked very well."

Ms. Dennison came over and started looking me over. "What have you been up to now?" she said.

"Just throwing out the garbage."

Before she could come up with a wise-ass comment, Mr. Decker helped Barstow through the door. He also was helped up on an examination table, and Decker removed his helmet.

"Why the delay, Mr. Barstow?"

"I had to pick up a souvenir."

"A what?"

"The weapon I knocked out of the alien's hand."

"You have that?"

"Yes, it got lodged in the corner."

"Very interesting, Mr. Barstow."

"I've got to get to engineering to examine the thing."

"No, you don't, Mr. Barstow," said Ms. Dennison. "In fact, you will be the first to be under the body scanner." I laughed, and Ms. Dennison commented, "You're next, Captain."

I shut up, and Ms. Kane rushed in. "Oh, Captain. Are you all right?"

"Yes, Ms. Kane. I'm just fine."

"Can I get you some water?"

"No, but I could use a drink."

"I can't do that, Captain. You are in the examination room."

"Then why don't we go to the galley?"

"Because you're up next for a full-body scan," said Ms.

Dennison from across the room.

"Let's get you out of that suit," said Ms. Kane.

She assisted me in getting my pressure suit off, and I felt a little woozy. "Are you sure you're all right?"

"No, I'm not all right. I just had a fight with a monster. I'm entitled to be a little shaken up."

Ms. Dennison walked in and had something to say. "Sit down and stop acting like a baby." She looked up and addressed Ms. Kane. "Does he have any marks?"

"A bruise on his left arm."

"How do you know that?" I asked. "You flinched when I touched you."

"Let's take a look," said Dennison. "Roll up his sleeve," she said to Ms. Kane.

"Ouch, that must've hurt," she continued. "He did that through a pressure suit?"

"Yes. He was very strong."

"Come over here," she said as she stepped over to a strange-looking instrument. "Put your arm in here."

"What is this?"

"I'm not sure. It came with our new installations."

"You're going to risk my health on a newly installed unit."

"When are you going to trust your medical staff? Now put your arm in there."

I did as she said and almost immediately felt relief. "That feels great."

"Now see there? I told you. Take your arm out."

I followed her instructions and saw no sign of a bruise. "That's magic."

"No, Captain. It's technology."

"Ms. Kane. If you will finish with this baby, I'll get back to Mr.

Barstow."

"I don't think she likes me," I said.

"You'd be wrong, sir."

"Maybe I'm just too sensitive."

"You'd be wrong again, sir."

"You've been hanging around Ms. Dennison too much."

"We both work in sick bay."

"That must be it."

CHAPTER 50

After my visit to sick bay, I returned to the bridge. "Mr. Basser," I said. "What's it like being back to work?"

"It's great, Captain."

"How's everything, Captain?" asked Mr. Anderson. "You did a very good job, Mr. Anderson. Thank you."

"I'm just glad that everything worked out, Captain."

"What was it like, Captain? Wrestling with that thing," asked Mr. Basser.

"Very scary and physically exhausting. The alien was incredibly strong. He grabbed me by the arm."

"How did you get away, Captain?"

"Mr. Barstow jabbed him in the arm with a hammer."

"That's just unbelievable."

"No, Mr. Basser," I replied. "Growing a new leg, that's unbelievable. That's enough questions for now. Besides, we will all be meeting in the conference room in two hours."

I told Mr. Barstow to retire to his quarters for some rest, and I did the same myself. As much as I was behind in my journal, I was too tired to do anything but sleep.

My wake-up call left me with little time to tidy up, so once again, a splash of water had to do. A surprising lack of corridor activity made me wonder what was going on. I was the last to arrive. That was what was going on. Upon entering the conference room, I was greeted by an applause from everyone. Mr. Barstow shook my hand and escorted me to the head of the table. Everyone was generous with the display of appreciation, and it lasted for almost a minute.

"Please. Everyone, settle down," I said. "We've got a lot to cover, and I want to check in with everyone."

"We hear that you kicked some alien butt, Captain," came a voice from the back of the room.

Everybody applauded and cheered again until I put out my hand for order. "I'm not sure I'd put it that way, but you obviously would, Mr. Decker."

"Yes, sir," said Decker, and a few chuckles could be heard.

"Let me set the record straight. It was Mr. Barstow that struck the blow that sent the alien on his way. Now I think we should go over some things that we know about the recent excitement. The exact classification of the creature is isomorph, which, I'm told, has something to do with its crystalline structure. Whatever it is, it is the scariest thing I have ever seen. I'm glad it's gone, and now maybe things can get back to normal."

"What's normal aboard a starship?"

"Not an isomorph, Mr. Decker. We also have an alien life-form aboard that has had a very positive effect, if you haven't heard. We have all been infected, for lack of a better term. It's why we have all gone through the body scans, and our research staff is working on it. One of the major benefits of the substance under study is the ability to enable us to grow severed limbs back. Mr. Basser has done just that, and if you would like to know what that's like, then you should get together with him in his off time."

"Are there any negative effects of the substance?" asked Mr. Langer.

"I'll let Ms. Kane speak about that."

She stood and addressed Mr. Langer's question. "We are dealing with a substance that is totally unique to our experience, so we can't assume anything. We have, however, detected no negative effects."

"Thank you, Ms. Kane. If there are no further questions, this meeting is adjourned."

As everyone was leaving, Ms. Grason stepped forward with Archimedes at her side.

"Captain. You seem to be the man of the hour again."

"It's just all in a day's work."

"Don't be so modest. It's nice to know we are in such good hands." Archimedes made a sound that could only be interpreted as approval.

"What can I do for you, Ms. Grason?"

"We have all been under a great deal of pressure as of late, and I was wondering if it might be time for another party."

"I think that is an excellent idea, and I think you should be in charge of setting it up. We do have a few things to take care of, so why don't you wait for my word before you proceed with the festivities?"

"I will wait for your word, Captain." Ms. Grason left the bridge, and I took my seat at the helm.

"It's good to see you back on the bridge, Captain," said Mr. Anderson.

"It's good to be back. Can I get you a cup of my special blend?"

"That would be just great, Captain."

I made two cups and returned to the master console. "Mr. Anderson, we haven't had much time to talk, so maybe we should take this time to catch up. What do you think of the mission so far?"

"It's scary as hell."

"I couldn't have put it better myself."

"I knew we would see some amazing things, but this is approaching overload."

"We have all had major responsibilities that have demanded us to go beyond the call of duty, and no one has met that more than you, Mr. Anderson. We all need a break, but that may not be forthcoming. I am depending on you for your continued performance level, and I believe you will be up to the challenge."

"Thank you, Captain. I will do everything I can to execute my duty to the best of my ability."

"I thank you for that, Mr. Anderson. Now let's both shut up and enjoy this great coffee."

For the next few hours, Mr. Anderson and I had an enlightening and most enjoyable conversation. When Mr. Blake came in to relieve Mr. Anderson, I left the bridge myself. I was hoping to get some rest, but I could see Ms. Kane coming down the corridor.

CHAPTER 51

I was not anxious to find out what she had to say, but I had no choice. "I must speak to you, Captain."

"Yes, Ms. Kane. What's on your mind?"

"Can we talk some place private?"

"Let's go back to the conference room."

She said nothing on our way back, and that suggested something serious. We came to the conference room, and she sat down immediately. I offered her coffee, which she declined, and I had just finished a cup, so I sat down to get to business.

"What is so urgent, Ms. Kane?"

"It concerns Mr. Basser's ability to grow body parts. A scan of cell fragments from the isomorph has revealed an element in common with Mr. Basser. We believe that the amorphous substance has given this ability to both the isomorph and Mr. Basser."

"Does everyone on board have that capability?"

"I'm not going to ask anyone to cut something off to find out."

"So where are we going with this?"

"When we brought the isomorph aboard, we thought it was dead. Maybe none of them were dead."

"Don't forget, Ms. Kane. We blew up the ship."

"There may be another ship."

"Ms. Kane. The ship looked like it might have been floating for centuries. We can't tell how long ago that battle took place. There are a lot of questions that may never be answered, but we can't drive ourselves crazy speculating about all of the possible ramifications. I suggest that you take a break and relax. In fact, get together with Ms. Grason and help her plan our next dinner party. We all need to relax and enjoy our present state of peace. Relax, Ms. Kane. You've earned it."

She left, but I wasn't sure that I got my point across. I retired to my quarters and made a few entries in my journal. Because of the recent flurry of activity, I had a lot of entries to make. It was good to know that the mission was coming to a close. Major scientific breakthroughs were made, and our discovery of extraordinary new life-forms would have a major impact on Earth. I finished my work and for some reason, had a great night's sleep.

At the start of my next shift, I found myself whistling on my way to the bridge when I spied Mr. Clark coming my way. "I hear we have a party coming up, Captain."

"Check with me about that."

I stepped onto the bridge where Mr. Blake was at the helm. "Good morning, Captain."

"Anything to report?"

"Smooth sailing, sir."

"Isn't Mr. Basser supposed to start his shift?"

"He's in the conference room getting some coffee."

"I hope he's getting some for me."

"Yes, Captain," said Mr. Basser as he entered. "Thank you, Mr. Basser. How do you feel?"

"I feel better than ever," he said exuberantly.

"I would like to make a slight course correction, Mr. Blake," I said. I sat down at my station and entered new coordinates. "Keep us there, Mr. Blake."

Mr. Barstow entered and came directly to me. "We must go to lab 2, Captain."

"I have something to take care of just now, so if you will go to lab 2, I will join you in a few minutes."

"No, Captain. This is too important. We must go now."

"Very well, Mr. Barstow, after you. The bridge is all yours, Mr. Blake."

I had never seen him so demonstrative. It certainly must be important, I thought to myself. The lab door opened, and I saw Ms.

Kane hard at work. She noticed me and called everyone's attention. "Let's all get to the subject at hand," she said. Ms. Dennison and Ms. Grason were there also. It must have been more serious than I thought. Ms. Kane continued, "All of us on the research staff agree that we have a serious situation on our hands. Mr. Barstow has brought this mater to our attention, and we feel it should be handled with some urgency. I feel Mr. Barstow should explain."

For a moment, he paused, as if reluctant to give bad news, then he spoke, "Ladies, Captain. For some time now, we have been plagued by a number of events that have more than just inconvenienced us. In addition to a substance that we all have been infected with, our recent visitor has given us much to be concerned about."

"Mr. Barstow. I don't mean to interrupt. But Ms. Kane and I had a discussion about this, and I reminded her that we blew up the alien ship."

"That may be so, but we have more evidence."

"Evidence of what?" I said.

"Do you remember when we were helped back into the ship by Mr. Brock and Mr. Decker?"

"Yes, I do."

"You lost track of me because I went back to retrieve the weapon that I knocked out of the alien's hand. It became lodged in the corner, and I have been studying it ever since."

"Have you been able to figure it out?" I asked. "Only how it operates, not why it operates."

"What kind of weapon is it?"

"A quantum weapon."

"What is a quantum weapon?"

"That's hard to explain."

"Give it your best shot."

"Very well, Captain. A conventional weapon works by physical damage from a projectile that pushes matter out of the way as it passes through material. More often, the wound is not fatal, but the victim bleeds to death. A quantum weapon, when set for carbon-based life, will totally disintegrate all organic life. It must be a direct hit for it cannot penetrate even the thinnest cloth."

"I wondered why it never fired the weapon."

"It would have done no good. We were totally covered with our pressure suits."

"Is that it?"

"No, Captain. The weapon can be set for almost any substance, which is why it got out of lab 2 so easily. That there may be other isomorphs in the area is a possibility."

"What are you suggesting, Mr. Barstow?"

"I would advise that we inform the military and get some ships out here to investigate."

"What is the best course of action?"

"We think it would be advisable to get to the communication project as soon as possible."

"You don't have to tell me twice. Let's get down to communications immediately."

CHAPTER 52

I returned to the bridge and sat down for a moment. "Are you all right, Captain?"

"Yes, Mr. Blake." I replied. "I want to change our coordinates." I entered the data at my station and stepped back. "Mr. Blake. If you will enter those coordinates into the navi-computer."

"Yes, sir."

"I will be back later to reset to a new set course changes."

I joined Ms. Yates and Mr. Barstow in communications. I was all business in my attitude for I was disturbed by what I heard from my research staff. Ms. Yates was keyed up and ready for the spotlight. "I'm about through coding, Captain, so if you will excuse me for a moment."

"Take your time, Ms. Yates."

Mr. Barstow was entering data also. I felt somewhat useless, but I was curious to see the process. "There," said Ms. Yates. "I'm ready, Mr. Barstow."

"Ms. Yates. This process is an exercise in frequency synchronization. We will not know if we have established a communication link until a series of operation signals are coordinated. A series of filters must be engaged to assure wave-shape integrity. Despite the fact that a digital control system is in place to operate the system, it is analog coordination that we must achieve. Sound is made up of analog signals.

"The containment matrix must be operated with a fixed-frequency, modulated digital signal. High-speed multifunction filters are used to keep the original audio signal free of interference. When you receive the carrier signal, the blue light will come on, followed by an audio tone that will turn off when the computer detects a perfect sign wave. When the audio signal disappears, shut down the link for you have established synchronization."

"But, sir," said Ms. Yates. "If this pair is perfectly in sync, why is synchronization necessary?"

"Electronically induced signals produce some degree of distortion of signals. The synchronization applies more to the response curves of the electronics both sending and receiving." Mr. Barstow paused for a sip of coffee. "Video signals are actually easier since they do not require such accurate signals, and our video-enhancement capabilities are so advanced. Are we ready to go, Ms. Yates?"

"Yes, Mr. Barstow."

"Watch your screen. In about ten seconds, a checklist will appear. The first space is labeled IRCS. This stands for Instant Response Communication System. There are three spaces to the right. In the first space, set and punch in. A running sequence of numbers will appear. When it stops, punch in. Go to the next space and punch in. Another running sequence will appear, and again punch in when it stops. Repeat this procedure in the third space."

"Done, Mr. Barstow."

"Okay, now go to the next on the list. This should read matrix control. Punch it in. There are a number of variables that apply to the interface of the matrix. Its power requirement is just three volts. Another power input is variable from one to three volts and is controlled by the same signal that controls the digital to analog converter."

I left Mr. Barstow to his explanation and made some coffee in the conference room. I sat down at the table for a moment to collect my thoughts. It had been a long time since I had a moment to myself, so I placed my face in my hands and tried to think of the accomplishments of this mission. Despite the great things we had accomplished, we might be known as the bearer of bad tidings. I lifted my head out of my hands and saw Ms. Cooper sitting across from me.

"Oh! Ms. Cooper, you startled me."

"I didn't mean to, Captain. You just seemed so tired I did not want to disturb you."

Ms. Grason suddenly appeared with a cup of coffee. "Yes, Captain. You've been under a lot of strain lately." She placed a cup in front of me and sat down beside me.

"I must say, this is an unexpected pleasure. What brings you ladies to the conference room?"

"Oh, we just punched our time clock and came down here to the bar to have a beer and pick up guys," said Ms. Grason. "They have no beer, but the guys are nice. Do you come here often?"

"More often than I care to remember. Now what is the real reason you're here?"

"We were just curious about the communication project," said Ms. Grason.

"Is it true that we can achieve instantaneous communication across vast distances?"

I waited about twenty seconds to answer. "Yes, Ms. Grason. It's the short-distance communications that take so much time."

"It is so wonderful to have a captain that moonlights as a stand-up comedian."

"But, Ms. Cooper," I replied, "I'm sitting down."

"Indeed you are, Captain. My mistake." Mr. Basser entered the room and went directly to the coffeepot.

"Mr. Basser," I said. "You certainly have acquired a taste for my secret blend."

"I've never tasted anything like it."

"Then by all means, help yourself."

"I will certainly do that."

"Ladies, Mr. Basser, I must attend to some business on the bridge, so if you will excuse me. I will be back for a second show in about thirty minutes."

"Thank you for the warning, Captain."

Mr. Anderson was running diagnostics on one of our systems as I set foot on the bridge. I did not seem to disturb him as I took my seat. I was impressed with his concentration until he acknowledged my presence. "Are we ready for the course change, Captain?"

"Yes, Mr. Anderson. Just let me enter the proper coordinates." I entered the data, and the navi-computer signaled an error. I entered again, and once again the navi-computer indicated an error.

"Captain. I would trust your judgment over the navi-computer."

"I appreciate your confidence, Mr. Anderson, but run a diagnostic on it after you finish what you are doing."

Mr. Basser returned to the bridge with his coffee and informed me that Mr. Barstow was waiting for me to return for the final procedures. I excused myself and headed to communications. Ms. Yates was all excited as I entered communications. She was also a bit nervous as Mr. Barstow gave her final instructions.

"It will take about two minutes to complete the process," said Barstow. "When the display light comes on, we should have successfully completed the process."

CHAPTER 53

We waited anxiously for what seemed a long time while Barstow continued his dissertation. It just didn't seem like it was going to happen, and then the light went on. Ms. Yates screamed out with excitement as if she had won the lottery. Mr. Barstow asked her to download all of our information on the isomorph.

Mr. Clark stopped as he was passing by. "I guess you finished the communication project."

"That would be safe to assume, Mr. Clark." I looked down the corridor and saw Mr. Basser. "What, Mr. Basser, more coffee?"

"No, sir. You are needed on the bridge immediately."

I headed out the door with Mr. Barstow, Mr. Clark, and Mr. Blake, who had just reported for duty. Mr. Anderson was still running diagnostics on the navi-computer when we arrived.

"What's so important, Mr. Blake?"

"It's an alien ship, Captain. It just came around the cloud."

"They must have seen us," said Mr. Basser.

"Not necessarily. Don't forget we have the corona killer," said Mr. Anderson.

"They are headed out of the dense corona area. We'll be in sight soon," said Mr. Blake.

"Then get us into that cloud," I said. "Main thrust engines, Mr. Blake."

Even with the IGS system operating, the acceleration caused us to stagger a bit. There were no major inertial consequences though, and we were soon inside the cloud. "Shut down all external lights and communication devices. Prepare for silent running. Mr. Barstow, shut down any nonessential systems. What about our communication project?" I said.

"Different principal. It won't be necessary to shut it down," said

Barstow.

"Slow us down, Mr. Blake," I said. "We don't want to come out the other side."

The entire crew was filing in and asking what was going on. I instructed them to join Mr. Barstow in the conference room. "Keep us in this fog, Mr. Blake, and stay alert."

Everybody found seats at the conference room table, and I took my place.

"I know you all want an explanation for what just happened. Just moments ago, we saw an alien ship emerging from the other side of the cloud. I took immediate action and ordered the firing of our main thrust engines to propel us into the cloud to avoid being seen. We shut down all nonessential systems to avoid being detected. The ship appears to be a heavily armed warship like the one we encountered previously. Are there any questions?"

"Can't they find us through the clouds?" asked Mr. Decker. "We can't be sure, but we hope not."

"Do we know if there are any consequences to entering the cloud?"

"No, Ms. Cooper, but we do know that we would be annihilated if we were discovered."

"How long do we stay here?" asked Mr. Langer.

"We can't be sure, Mr. Langer. We'll be busy though."

I answered a few more questions and then paused. "The following crew members remain seated: Mr. Barstow, Ms. Kane, Lorin and Lanie Rodgers, Mr. Matson, and Mr. Palasco. Everyone else, return to your stations and wait for instructions."

"Captain?"

"Yes, Ms. Yates?"

"How do we communicate?"

"You and Mr. Clark will be in charge of getting messages on

paper from one station to another." I stood and addressed the crew members that remained. "We will need another bomb. It must be twice the power of the first one we set off and must have a rather complex computer detonator. Mr. Palasco, can you work with the Rodgers sisters and Mr. Barstow on this project?"

"Yes, sir."

"Mr. Barstow, do you have any idea why they didn't see us when we saw them coming from behind the cloud?"

"Don't forget, Captain. We have the corona killer."

"But, Captain," said Ms. Grason, "even if they didn't see us directly, the isomorphs have thermal vision. They should see our thruster trail."

"They may have, which is why we must walk softly down the corridors. Unfortunately, we can't seem to track them, and Mr. Barstow will answer that."

"They may have a cloaking technology," said Barstow. "Then they know we are in here," said Ms. Kane.

"That is true," said Barstow. "But they won't fire into the cloud."

"Why is that?" asked Ms. Kane."

"They are warriors and like to see the kill. Besides, they don't know what firing into the cloud might produce. There is one other possibility."

"What is that, Mr. Barstow?" I asked.

"They might want to capture us. That would give them information about our system, and even accidental discovery of star maps would reveal Earth's position."

Silence followed Mr. Barstow's remarks as the gravity of the situation settled in. Our decisions now would determine the survival of the Mobius. "Mr. Matson, I want you to return to engineering and prepare for the fastest acceleration we can possibly attain."

"Yes, sir," he said as he got up from his chair and walked out of the conference room.

"Mr. Brock," I said. "I want you to come up with some method of tracking the alien ship."

"I'm already thinking along those lines."

"Ms. Grason. I want you to work with Mr. Brock on this."

"Yes, Captain," she said, and they both headed out the door.

"Ms. Kane, Mr. Barstow. I need you both on the bridge. And Ms. Kane, I need to speak to Ms. Dennison, so if you will please go down to lab 2 and bring her up here."

"Yes, Captain," she said as she quickly got up and left the bridge. I turned to Barstow for advice and was not disappointed. "Captain. I must tell you that your decision to propel us into the cloud saved our lives. Our contact with the isomorphs has convinced me that they pose an inevitable threat to Earth, and we must prepare for an invasion. There are a lot of things that we don't know about them, but several things are clear. They are a species that is geared to conquest, they have quantum technology, and Earth is well suited to their environmental requirements."

"Mr. Barstow, do they seem to you like the kind of creature that would have developed quantum technology?"

"You noticed that, didn't you?"

"They just don't seem to be rational, intelligent beings that are capable of advanced scientific research."

"They may be a slave race used to conquer new worlds for a more advanced species."

"Is that what you believe?"

"It's just a theory."

I took a moment out and stepped to the door. "Ms. Yates," I called and stepped back.

In a few seconds, she appeared at the door. "Yes, Captain?"

"How is the project going?"

"Everything is just perfect. The information feed is rolling along."

"Where is Mr. Clark?"

"He's running messages, sir."

"When you see him, tell him to report to the bridge."

"Yes, sir," she replied and returned to the communications room.

CHAPTER 54

I turned my attention back to Mr. Barstow and saw him looking at the white screen. He seemed hypnotized by the blank canvas in front of him. He spoke while still looking at the screen, "It's hard to believe we're in space." That sounded a bit too metaphysical for Barstow.

Mr. Clark appeared at the door, out of breath. "You don't need to run, Mr. Clark," I said. "Well, sir…"

"Sit down and catch your breath." I made a cup of coffee, and in a minute, he was breathing normally. "A brisk walk will do, Mr. Clark."

"Yes, sir."

He continued on his way, and I took my seat at the helm for a few moments. Ms. Kane entered the bridge with Ms. Dennison.

"We need to speak to you, Captain."

"I have a few questions I need answers to also."

We left the bridge and settled down in the conference room. They both sat down, and I started my questioning. "Ms. Kane, both you and Ms. Dennison worked on the green substance that was responsible for Mr. Basser's new leg and on samples from the isomorph. Did either of you perform experiments where the two were combined?"

"Yes, sir, we both did," Ms. Kane replied. "And what were the results?"

"Inconclusive. We both feel that a longer incubation period is required."

"Are they still incubating in the lab?"

"Yes."

"I want you to destroy the sample."

"Yes, Captain. We feel that under the circumstances, that is the best procedure."

"Ms. Dennison, you are assigned the task of disposing with anything that the science staff considers a possible hazard. I want you to take care of that immediately."

"Consider it done," she said and left the conference room.

Ms. Kane and I returned the bridge with Mr. Barstow. We conferred with him to make sure that we were doing everything right. He agreed that the emergency at hand precluded any time to devote to useful research. Mr. Clark reported in, and this time, he was not out of breath. "Captain. The Rodgers twins asked me to tell you that they do not have enough chemicals for another binary bomb."

"Thank you, Mr. Clark," I said. "And you're not out of breath."

"No, sir," he said with a smile.

"I think I can help with this problem," said Barstow.

"Very well, Mr. Barstow. Get down there and do what you can for the Rodgers twins."

He left the bridge and gave me the opportunity to ask Ms. Kane some important questions. "Ms. Kane. You have the most direct research experience with the isomorph. Are there any genetic weaknesses that could be exploited in the development of weapons?"

"I wish I could say yes, Captain, but nothing that I discovered revealed anything specifically exploitable."

"You do understand why I ordered the destruction of the samples."

"Yes, sir. If we have a potentially dangerous organic substance, crew safety takes priority. And, Captain, I have no problem with that."

"I just didn't want you to think that I was superseding your authority."

"Not at all, Captain. I was going to order it myself. I might add that it's great that we have a captain with a scientific background to keep us on our toes."

"I try to do what I can, Ms. Kane."

"It does not go unappreciated, I assure you."

"We have gone through so much lately that I imagine that everybody is about down to their last nerve, and still I have an intact crew. I even pat myself on the back every morning for not coming unglued. We all have been tested time and time again and have put ourselves in danger for the mission. As you can see, Ms. Kane, I'm babbling on when I should just shut up. We all need a vacation."

She started laughing, and I soon joined in uncontrollably. We must have laughed for at least a minute. Then we gathered ourselves just before Ms. Yates entered.

"Captain?"

"Yes, Ms. Yates."

"Do you know where Mr. Barstow is?"

"Yes. He's with the Rodgers twins. Is there anything wrong?"

"No, Captain. I just had a few questions about the download."

"He's due back any second. I'll tell him to drop by communications." She turned and went back to her station. Ms. Kane watched her as she left.

"She really didn't like that."

"Like what?"

"Seeing us together."

"What are you saying?"

"Oh, Captain. She's crazy about you."

"I just thought she was crazy."

Mr. Barstow returned to the bridge and made himself a cup of coffee. "Anyone care for some coffee?"

"I don't believe so, Mr. Barstow. How did things go in lab 2?"

"We now have a bomb that is at least two times more powerful than the first."

"Were they pleased with your suggestions?"

"I believe so, Captain. They looked at each other and smiled."

"Oh. The smile. They liked your idea."

"Is that their sign of approval?"

"Big time. Oh, I almost forgot, Ms. Yates would like you to join her in communications." I gave Mr. Barstow a chance to sip his coffee before I questioned him further. "Did you go by engineering?"

"Yes. Everyone is working hard on to increase the power output of our main thrusters."

"And Mr. Brock?"

"I think he's got something in mind. Give him another hour."

"You should check in with Ms. Yates before we start our strategy session." He got up immediately and left to take care of his business in communications.

I sat back for a moment to collect my thoughts. Ms. Kane poured me another cup and sat down beside me. "We are in some serious trouble, aren't we?"

"I'm afraid so. It's going to take a concerted effort to get out of this situation."

She took a sip of coffee and put the cup down. The look on her face was not what I wanted to see.

"I think we should discuss some strategy," I said.

"I believe that is more your area of expertise, Captain."

"We are going to need every thought we can get through this operation, and I have full confidence in the combined abilities of a talented crew."

CHAPTER 55

Barstow returned, and we got down to mapping out a plan of action. "If we can detect their position," said Barstow, "I would like to send out a bomb in one of our probes and explode it at a distance to attract their attention. When they start chasing the light, we fire our main thrusters and head in the opposite direction."

"Why don't we just blow up the ship like we did with the other one?" I asked.

"That was just a derelict ship. We don't know what a fully functional ship can handle. They are, after all, a warship with quantum technology. I assure you that they can easily handle anything we can throw at them. We do not want to be involved in a confrontation with them."

Barstow stood and took a sip of coffee before he spoke. "We have in our possession a quantum manipulator that could spawn a new technology. Certain unfamiliar material used in the device has made examination arduous process. We really need services like the Quantum Lab, outside Juneau, Alaska provides. One of the most unique features of the weapon we secured is its ability to target a specific material. It can disintegrate a steel plate without harming a silk scarf on the other side. That could be a process that only an isomorph could perform."

"When Mr. Brock is free," I said, "I would like you to work with him on this."

"I intend to do just that."

We sat quietly for a minute, trying to think of anything we might have missed. The silence was quickly broken though by the Rodgers twins and Mr. Palasco.

"My goodness," I said. "That is bigger than the one you made before."

"Yes, sir," they said in unison. "That will blow the hell out of them."

"Mr. Palasco, what have you devised?" I asked. He stepped forward to explain. "The last bomb was designed to explode after it had a slight impact of placement followed by the mixing of the binary chemicals and an electronic detonation from a control box. This bomb requires only one signal that is accessible from the main console and will set off the bomb and at the same time, trigger our main thruster engines. The bomb should provide ample distraction while our main thrust engines set us off in the opposite direction. One factor is critical though. We must be able to determine exactly where the alien ship is located."

"Can you set this up on the console?"

"Yes, sir. It will take about a minute."

"Then get to it and get the bomb installed into the probe, Mr. Barstow."

The twins and Barstow left the bridge. I looked at Ms. Kane as she sat down at the console and watched Mr. Palazzo entering data. "Well, we seem to be prepared for the escape part of the plan. Now all we are missing is the tracking part of the plan."

"He should be here shortly, Captain" said Ms. Kane. Mr. Palasco finished and left the bridge to help with the installation of the bomb. I sat down and leaned back to rest for a minute.

As we waited, Mr. Clark showed up at the door and made an announcement. "Mr. Brock is on his way."

"Thank you, Mr. Clark," I said. "Come join us."

We continued to wait for Mr. Brock to arrive, and very soon, he did. Ms. Grason was with him, and they seemed anxious to get to it. "I assume you have found a way to track the alien ship," I said.

Both Ms. Grason and Mr. Brock nodded. "We think we have come up with an interesting solution," said Mr. Brock.

Mr. Barstow returned to the bridge, and we all retired to the conference room. Everyone took a seat except for Mr. Brock. He looked around the room and then spoke, "For the last fifty years, scanning techniques have continued to develop. Our present state of the art uses the detectable disturbance of dark matter. Nothing we have come up with in the last ten years is better for tracking short-distance objects. Unfortunately, the ship we are trying to track has a cloaking device that renders our dark-matter tracking system useless. Before laser tracking, airports on Earth used a system called radar to track short-distance air traffic. This is what we're trying to do, so radar seems to be the best answer. Looking back through our archives, I have managed to design a circuit that can be installed, which will give us a radar capability."

"How do we install it?"

"There is an input access port in the main console that will accommodate the circuit."

"What circuit?"

"Oh. The one in my pocket." He pulled out a small circuit module that fit in the palm of his hand. "We just have to remove this side panel and plug it in."

"Well, let's get to it," I said. "Mr. Clark, go down to engineering and tell Mr. Matson that we are ready to go and then get back here. Mr. Brock, you will handle the radar system. Mr. Blake, you will stay at the navigator's console. Mr. Barstow, you will handle the launch and detonation of the bomb. Ms. Kane, Ms. Grason, you will remain to observe and advise. Mr. Brock, engage the radar." He entered the codes, and in just a second, the screen lit up with an operating system."

"We have a reading, Captain."

"Excellent, Mr. Brock," I said as I looked around and saw a smile on everyone's face.

"Where are they located?"

"Behind us, Captain."

"Mr. Blake. Engage our ion engines at 5 percent and enter the coordinates I just punched in." Everyone remained still for a few seconds. The silence was deafening as we waited for something. "Have they detected our movement, Mr. Brock?"

"I don't think so, Captain."

"That's great. We know where they are. Get us turned to Earth, Mr. Blake, and stand by."

Our course change would take about twenty minutes to complete, and while the adjustment was taking place, our strategy was being mapped out.

"Mr. Clark. I want you to go to engineering and tell Mr. Matson to get ready for a full-thrust firing."

Mr. Barstow pulled me aside and made a suggestion. "Captain. We do not know what effect our full-thrust departure will have on the cloud, so I would suggest that we detonate the bomb first and wait for a reaction from the alien ship and then apply full thrust after we leave the cloud. This will make sure that we still have the cloud between the alien ship and us."

"That does sound like a better idea," I agreed. "So I will wait for your signal to fire the thrust engines."

"Yes, sir."

I returned to the console and sat next to Mr. Blake. "Captain," he said. "This is the slowest turn I have ever executed."

"Keep her turning, Mr. Blake"

CHAPTER 56

The atmosphere was electric, and everyone was on edge. I wanted to get this thing going, so I looked to Mr. Brock and asked again, "Where are they now?"

"Moving slowly."

It was difficult to stay calm, and my crew seemed to be doing a better job than me. My immediate concern was to stay composed, so I took a deep breath and settled back. Ms. Kane handed me a cup of coffee. "Oh, thank you, Ms. Kane." A sip was all I needed to regain prospective.

"You looked like you needed it, Captain."

"You're right, Ms. Kane. Oh, I just remembered something I've got to do." I left the bridge and headed to engineering to speak to Mr. Matson. I was not sure what I would find, but my curiosity would soon be satisfied as Mr. Matson greeted me in high spirits.

"Captain. I'm glad that you've managed to find some time amidst all the chaos."

"I see some new equipment."

"Yes, Captain. That package just in front of you is ready for installation."

"And what is this, Mr. Matson? It looks like a bassinet."

"It is my baby."

"And what can your baby do?"

"It will increase the efficiency of our main thrust engines by 10 percent."

"How long will it take to install?"

"About ten minutes."

"You don't mind if I time you?"

"Not at all, Captain. Mr. Decker, give me a hand with this."

They picked up the device and placed it gently inside the main engine cover. About six lockdown bolts and a computer interface connection, the job was finished.

"Only five minutes and ten seconds. Very good, Mr. Matson. We're ready now?"

"As we'll ever be."

"Then I should get back to the bridge."

I left engineering with renewed confidence and saw Mr. Clark headed my way. "Captain. You're needed on the bridge."

"Then let's get going."

"This is a tough one, isn't it, Captain?"

"The toughest, but if everybody does their part, we'll get through this."

On the bridge, Mr. Brock was anxious to inform me that the alien ship was on the move. "They have increased their speed and are about ten minutes away from optimum angle."

"Everyone, stay alert and keep your eyes open," I said "We are about to get to the active stage of this operation. Mr. Blake, are we in escape position?"

"Yes, Captain."

"Mr. Barstow. Launch our probe."

"Increase our speed, Mr. Blake, and pull us out of the cloud."

Suddenly, we could see the sky. We were all relieved to see stars again. The job was still at hand though, and everyone seemed to tense up again. "Mr. Brock. Where is the alien ship?"

"It's coming into position, sir."

"Stand by, Mr. Barstow."

"It's almost there...Now, Captain."

"Fire the probe, Mr. Barstow."

"Mr. Brock. They haven't changed course. Wait, yes, they have."

"Fire main thrust engines, Mr. Blake."

We lunged forward with great force, followed by another blast a few seconds later.

"What was that?"

"What I was afraid of," said Mr. Barstow.

"Yes, Mr. Barstow?"

"Our thrust engines incinerated the cloud."

"Then we have no cover?" said Ms. Kane.

"Yes, Ms. Kane, so we had better keep an eye on our backside. What about the radar, Mr. Brock?"

"Out of range, Captain."

"Let's power back up. Put all of our systems back online, and, everybody, back to your stations. Mr. Barstow, I want to see you in the conference room in five minutes. Mr. Blake, keep our present course and let me know if anything is following us. And, Ms. Kane, if you will go to the conference room also, I will be there in a few minutes." I headed to communications and found Ms. Yates busy at her station. "Have we finished downloading the primary information?"

"Yes, Captain."

"Then start downloading the current emergency information. I'll be in the conference room for the next hour or so."

I left communications and slowly made my way to the conference room. How would I open this meeting? Did we wipe out a life-form? What could I say? The answers to my question would not be long in coming as the door slid open and I entered the conference room. Mr. Barstow and Ms. Kane were involved in conversation until Ms. Kane noticed me.

"Captain," said Barstow. "We must discuss a few things about our escape procedure that obviously impacted you. Let me just say that no one could have predicted what happened. Ms. Kane and I worked with the substance as best as we could, and neither of us had the idea that this would happen. We thought we had enough distance to avoid the thermal impact. Maybe it was the new unit that Mr. Matson added just before we fired the main thrust engines. It was a compound miscalculation that was unavoidable."

"But to have completely wiped out a unique life-form," I said.

"We didn't," said Ms. Kane. "Our sampling tanks are filled with it. We have tons of the stuff."

"We do?"

"When we entered the cloud, our sampling tanks were wide open."

"But doesn't it disappear?"

"We've been meaning to schedule an update meeting, but things have been so busy we've had no opportunity," said Barstow.

"Well, how about now?"

"Very well," said Barstow. "The substance, when in its powder state, is without cognizance and only tends to cling to itself to create billowy clouds. When it contacts something, it becomes cognizant and is capable of penetrating and passing through inorganic material in search of organic material. It craves consciousness, and it identifies DNA and strives to improve its host. That's why your broken leg did not show up on scan. When it repaired Mr. Basser's leg, its only reference was the other leg, so it made an exact mirror image of the other leg."

"There is something called a threshold point," said Ms. Kane. "When the powder content of any organic material reaches a certain percentage of the total mass, no more powder will be absorbed. Inorganic material does not have a threshold point because it is just a transfer medium and cannot contain the powder in the long-term. We have discovered that a small amount of applied direct current will keep the powder contained. It was most likely that we were protected by the powder when the magnetic flux passed through us, and don't forget our ability to breathe under less-than-ideal atmospheric conditions."

"Ms. Kane," I said. "I am about to fall over from information overload. Could we please continue this after I get some sleep?"

"Of course, Captain. It's a lot to take in," she said sympathetically. "You should get some sleep also," I said. "And you, Mr. Barstow."

"I am still on duty, Captain," said Mr. Barstow

"No, you're not…I want the both of you to get some sleep. That's an order."

"Yes, sir," they both said

"I'll see you in eight hours and not a minute before…so get to your quarters and get some rest."

They both left, and I headed to my quarters. I would certainly not have any trouble getting to sleep. I just hoped it would be a restful sleep.

CHAPTER 57

My head was reeling, so when I reached my quarters, I fell asleep immediately. I had wonderful dreams of amazing images and color displays. Like before, the dreams were restful and exhilarating. I woke rested and ready to get to the bridge. On my way there, Mr. Clark stopped me in the corridor and directed me to the conference room.

"What's in the conference room, Mr. Clark?"

"Something I wanted to show you."

"This had better be important."

"I assure you it is, sir."

The entire crew was there and cheered as I entered the room. I was startled and at first speechless. Mr. Barstow stood though and spoke, "We have gathered here to express our appreciation for your skills, Captain. Your decisions have gotten us through some very trying experiences. Both your encouragement and support have enabled us all to perform at the peak of our ability, and because we have survived a number of dangerous situations, we would like to toast to you, Captain." Barstow raised his glass and, along with everyone else, took a drink. He then sat down and left me to speak for myself.

I looked around at everyone and said the first thing that came to my mind: "Any excuse to open a bottle of wine." They laughed and seemed to wait for my next line. "Where is my glass?"

"You're not supposed to toast yourself, Captain."

"Is that you again, Mr. Decker?"

"At your service, sir."

"I might have known that you would be aware of this one rule of etiquette. I am prepared to sidestep the rules of social behavior and say that drinks are on the house, or should I say, on the ship."

Without warning, a party broke out. Mr. Clark left to get more wine, and Ms. Yates bestowed her congratulations in the form of a hug. Everyone was mingling, so I expressed my appreciation to Mr. Brock. "Your radar system worked superbly, and we're all grateful for your contribution."

"Thank you, Captain. I'm certainly glad it worked."

"I assure you, we are all glad it worked," I said as I shook his hand. "I did not notice Ms. Kane."

"She's on the bridge with Mr. Blake, sir."

"I think I'll take a couple of glasses of wine to her and Mr. Blake." I filled two glasses and headed just up the corridor to the bridge.

"Ms. Kane, Mr. Blake. I have some liquid refreshments for you both."

"Thank you, Captain. That's very nice," she said.

"I would also like to get some answers to a few nagging questions I have."

"What questions, Captain?" she said and then took a sip of wine.

"It occurred to me that any species capable of producing quantum weapons might also have the ability to develop a hyperdrive propulsion system."

"That is a reasonable assumption. However, the computer resources necessary to manipulate just a single quark are enormous. To expect to control a power source capable of propelling ships at supper speed is not realistic. Besides, they are more interested in weapons development, and they know that quantum physics is more suited to that."

I accepted her explanation and returned to the party. I spoke to Mr. Anderson, and he was satisfied that we were not beginning to get followed. Mr. Barstow though was not too sure, and he took me aside to speak. "Captain. I am disturbed by the ease with which we escaped."

"Mr. Barstow, I would like to think that it was the result of some piloting skills of mine, the knowledge of our chemists, our inventor Mr. Brock, our engineering crew, our researchers, and I might add, your own coordinative and advisory talent. And yes, a bit of luck that gave us the ability to escape."

"Somehow I don't think we've seen the last of the isomorphs."

"That's why we are downloading all the information we have about them. I think we are doing everything we can regarding this situation. If there is any further action we can take, I'm listening."

"We are still studying the quantum weapon and reviewing all the other experiments, and we have months before we get back to the Stockton. We are also keeping a constant watch for anything on our tail. And I also checked with Ms. Kane, and she seems to think we made a clean getaway."

"She may be right, sir. It's probably just my natural skepticism."

"There is one question I wanted to ask."

"Yes, sir?"

"Do you think that the magnetic flux force came from the alien ship?"

"Most definitely."

"Do you think that the signal was sent out to kill carbon-based life-forms?"

"It is not unreasonable to come to that conclusion."

"Then you agree with Ms. Kane that we were protected because we were infected?"

"There is no better explanation."

I was concerned about the arsenal of weapons that the aliens had. If they were to find Earth, we would be in for the fight of our lives. It was fortunate that we had tons of the life-saving substance in our collection tanks. All of this new information was giving me a headache, so when Mr. Anderson suggested we go to the bar for another glass of wine, I was more than happy to join him. Mr. Basser was pouring, so I took the opportunity to ask some questions.

"How's the leg?"

"It's just fine, Captain," he said cheerfully. "You mean it feels normal?"

"Not exactly, sir."

"Then what do you mean?"

"Somehow it works better. Or they work better together. They tested me. I run faster now than when I ran track and field."

"Was there pain during the healing process?"

"You mean when my leg was oozing that green stuff?"

"Yes."

"No, sir. Quite the opposite. I was in a state of euphoria."

"Did you have vivid dreams of incredible shapes and colors?"

"How did you know that, sir?"

"I had them also." I looked up to see Ms. Dennison, so I excused myself and walked over to her table.

"Don't try to sneak up on me, Captain."

"Now why would I do that?"

"Just your devious nature. You can't help it."

"Dabbling in psychiatry now, Ms. Dennison?"

"Yes. In case this doctor thing doesn't work out."

"That's what I like: a woman that thinks about the future."

"If you have something to say to me, get to it."

"I think I like Ms. Kane better than you."

"I'm crushed. Now what do you want?"

"I spoke to Mr. Basser about dreams he had when he was healing. They were identical to my dreams."

236

"It doesn't mean a damn thing."

"Are you sure?"

"Of course, I'm sure. I'm the doctor."

CHAPTER 58

We were finally going home, and I felt like a weight had been taken off my shoulders. It seemed that the entire crew was more relaxed, and smiles were more abundant. My arrival on the bridge was greeted with a cheerful Mr. Anderson. We both had a cup of coffee to shake the lasting effects of an evening of wine. I took my place at the console, and for a while, we were exchanging anecdotes and stories about life at the academy.

I reminded Mr. Anderson of a course change that would avoid the asteroid field we encountered on the way out. Mr. Clark joined us, and we continued our friendly banter. Ms. Grason appeared at the door and asked me to join her in the conference room. I excused myself from the bridge and followed Ms. Grason to the conference room.

"Can I get you something, Ms. Grason?"

"No, thank you, Captain."

"Ms. Grason. What is it you wanted to speak to me about?"

"Captain. Archimedes found something on the outside of the ship."

"And what would that be, Ms. Grason?"

She pressed her personal communicator, and Archimedes entered the room. "Hi, Archie," I said. He returned my greeting with a cheerful-sounding series of tones. "What do you have for me, Archie?"

"Put it on the table," said Ms. Grason. Archie moved to the table and plopped down something that looked like a small turtle."

"My god, what is that?" I said. "I think it is a tracking device."

"Mr. Anderson," I said as I pressed my communicator button. "Find Barstow and get him up here."

I approached the device and looked closer. It did appear to be something that the aliens would make. Ms. Grason also came forward to get a better look. While we were examining the device, Mr. Barstow came in. "Mr. Barstow. Archimedes found this device attached to outside of our ship."

"It's a tracking device," he said rather quickly. "That's what Ms. Grason said."

"We've got to get rid of it"

"When did we acquire that?"

"They probably sent it through the cloud."

"Then why couldn't they track us?"

"They weren't sure it was attached to us or something else that might be in the cloud."

"Why didn't I think of that?"

"I have no idea, Captain."

"I did think of one thing. We are about to pass through the asteroid field that we encountered on the way here. I told Mr. Anderson to adjust our course to avoid the asteroid field. Mr. Barstow, go to the bridge and readjust our course. Ms. Grason and I will go down to engineering and see if we can come up with a delivery system."

Ms. Grason, Archimedes, and I headed to engineering. Mr. Matson greeted us at the door. "Mr. Brock is in his workshop, Captain."

"Word gets around," I said. Mr. Brock was sitting at his workbench. "We need your services again, Mr. Brock. It seems we've picked up an unwanted passenger," I said.

"And you want to attach it to one of the asteroids in the field we are approaching?" said Mr. Brock.

"You read my mind," I replied.

"May I see the device?" said Mr. Brock. "We do have a small probe that can attach to just about anything, and I believe I can adapt it to accommodate our stowaway."

"How long will it take to get everything ready for launch?"

"About ten minutes."

"Then get to it, Mr. Brock."

Archimedes moved to the workbench and placed the device in front of Mr. Brock.

"Come with me to the bridge, Ms. Grason, and you too, Archie." We stepped lively, and we arrived in record time. "Mr. Anderson, have you readjusted our course?" I said.

"What is our ETA?"

"About one hour."

"Give us a two-second blast on our main thrust engines."

"That should get us there in fifteen minutes, Captain," said Barstow. "Will the probe be ready?"

"With five minutes to spare, Mr. Barstow."

We started a countdown, and everyone sat down at their stations. Ms. Grason selected launch port 5 and prepared for loading. Mr. Anderson was setting controls for manual operation, and Mr. Barstow was overseeing the operation.

"We are approaching launch time, and there is no probe in the port, Captain," he said.

I pressed my personal communicator badge. "What's the problem, Mr. Brock?"

"I'm just about ready, Captain."

"Get a move on, Mr. Brock."

We waited for another minute, and still the launch port was empty. "Captain," said Mr. Barstow. "We have less than two minutes to launch."

"Mr. Brock. What's the problem?"

"Just a minute, Captain."

"We don't have a minute, Mr. Brock."

"Captain," said Ms. Grason. "The port is loaded."

"Standby for launch, Mr. Barstow." The launch display was at ten seconds, and without verbalizing the count, I said, "Launch, Mr. Barstow."

He pressed the launch button, and we could see the probe as it flew away. For the next few seconds, Barstow maneuvered the probe to attach itself to the largest of the asteroids.

"Mr. Anderson, I leave it to you to get us out of this area and back on course, and I would like to commend you, Ms. Grason, on your alertness. Oh yes, thank you too, Archimedes, for finding the tracking device." Archimedes made some pleasant noises that I think I was actually beginning to understand.

"Let them track that," said Mr. Anderson.

"I believe we can consider this operation a success," I said.

"Yes, sir," said Barstow "We do have another piece of business that we must take care of."

"And what is that, Mr. Barstow?"

"Ms. Dennison and Ms. Kane have scheduled a meeting to update you about the current stage of research."

"And when is this to take place?"

"They are waiting in the conference room right now."

"Then let's not keep them waiting. Mr. Anderson, the bridge is yours."

CHAPTER 59

I wondered what business we needed to take care of with such immediacy. I just kept my mouth shut and waited till I saw the whole research staff sitting at the conference table, stirring coffee in unison. Ms. Kane stepped forward and spoke first.

"Captain, we have a problem. When we first came in contact with the amorphous substance, it was hard to study because of its ability to combine with anything it touches. The medical potential of the substance seems to be nothing short of phenomenal. It has cured every malady in everyone on the ship including old injuries, the removal of scars, and even the regrowth of a severed leg. Although this promises to be the greatest discovery in the history of medicine, there is a problem. We have done extensive study of our DNA chain since we have been infected. Major changes have been discovered, and we don't have enough time to evaluate the long-term effects."

"We certainly know the short-term effects," I said.

"Yes, and they are far-reaching," said Ms. Kane. "It would be ideal to return to Earth with such a valuable cargo, but we have not had enough time to discover the long-term potential for negative effects on our DNA."

"And what if there aren't any?" I said.

"Elements of our DNA double helix have been changed in ways we do not understand, and the few months we have before we get back will not be enough time."

"Why have you chosen to assume that these changes are not, or will not, be of wondrous benefit to the future development of mankind?" I argued.

"It's not that we have chosen so much as we are exercising caution. You must admit, Captain, that we have experienced a number of highly unusual things, things that no one has ever seen before. We can't assume that there are wondrous things with such limited experience."

"But there are wondrous things that have revealed themselves," I continued.

"On the surface, that might be true, Captain, but such monumental reconstruction of our DNA is happening that we can't know what the future consequences are."

"But that is exactly what you are doing."

"You are not making this easy, Captain."

"Someone must defend the obvious. You stated that major changes have been made in our DNA. Are you saying that you are the one to determine that these benefits should not be afforded the people of Earth?"

"No, Captain. I do not want to be responsible for the plague of the Earth if it is discovered that DNA changes are not beneficial in the long run."

"What are you proposing, Ms. Kane?"

"I feel we should dump all of our collection tanks."

I got up, walked over to the coffeepot, poured a cup, and then returned to my seat. Nothing was said for several minutes. I don't think anyone knew what to say, so I broke the silence. "Do you all feel this way? How about you, Ms. Cooper?"

"I am not a medical professional, so I must defer to those who are."

"Ms. Dennison, how about you?"

"Ms. Kane's arguments make sense to me."

"And you, Mr. Barstow?"

"Because the consequences are permanent and unforgiving, I must agree with the others."

"Well then, I think that you should handle the details, Mr.

Barstow. I want no part of this operation." I stood and left the conference room. My walk back to the bridge was somber, and fortunately I passed no one. When I sat down at my station, Mr. Anderson offered to freshen up my coffee. "Make it a double." I said.

"Pardon me, sir?"

"Never mind...So, Mr. Anderson, tell me about your days at the academy."

"Well, sir, there was this one time that..."

Mr. Barstow walked into the room and tapped me on the shoulder. I excused myself, and we went to the conference area.

"Captain, I must speak to you about this situation."

"What situation?"

"The dumping of the substance."

"Go ahead."

"You know we can't do that."

"I will not take part in something like this."

"But, Captain, you know we can't dump anything without your permission."

"Then let it be known that against my better judgment, I agreed to do this only because of the recommendations of my entire research staff."

"This is quite unprecedented, Captain, but we can arrange for your feelings to be specifically documented."

"Then see to it, Mr. Barstow. And now I am going to return to my station." I returned to the bridge again and started my conversation with Mr. Anderson.

"Are you all right, Captain?"

"I have never been better."

"I don't think I have ever seen you so interested in stories about academy adventures."

"Well, Mr. Anderson, the mission is winding down, and our duties have diminished. So there is more time for socializing."

For the next hour, I listened to Mr. Anderson's adventures, and although many of them were surprisingly amusing, I could not shake that feeling.

Ms. Kane appeared and sat down beside me. "Captain. Can we have a minute in the conference room?"

"Yes, Ms. Kane."

We stood up and walked to the conference room in silence. The research staff was still there, and I fought the urge to walk out.

"Captain," said Ms. Kane. "We don't feel that we represented our position clearly."

"Then clarify, Ms. Kane."

"As a medical professional, I would give anything if we could just haul the substance back to Earth and wipe out diseases. All of us here are as disappointed as you are, but I would be acting irresponsibly if I didn't consider the consequences of taking a substance to Earth that had a yet-to-be-discovered potential for disaster."

"Captain," said Mr. Barstow. "As the science officer of the ship, I must consider the consequences of all my recommendations.

I have consulted with Ms. Kane, one of the most experienced researchers in the field of DNA. Ms. Dennison, our chief medical officer, agrees with Ms. Kane. Add to this my own science expertise, I must recommend dumping a potentially hazardous substance."

"What about all the potential benefits?" I said.

"Captain. Do you remember when you had the bad dreams?"

"Yes, Mr. Barstow."

"And then you had some fantastic visual dreams that Mr. Basser also had. You suggested to Ms. Dennison that mutual contamination could be the cause."

"Yes, but she assured me that was not the case," I said. "And what if you are right?" he replied with emphasis. "What do you mean?"

"Any substance that can exercise such mind control must be considered a threat to you."

For the next five minutes, not a single word was said by anyone. I stood and walked to the coffeepot and filled my cup. Then I paced around the room for another two minutes The ladies seemed concerned with my response. Mr. Barstow was relaxed and calm as he sipped his coffee. He knew that there was no logical argument that I could present that would support my point of view. I decided to pace just a little longer to keep them in suspense. I turned to them with a stern look on my face and said, "I'm afraid that I must insist that..."—I paused again for a few seconds—"that we go ahead and dump the substance."

They all stood with smiling faces, even Barstow, and shook my hand. The tension quickly vanished, and everyone was relieved enough to head to the coffeepot for one more cup.

"Captain," said Ms. Kane. "I thought you were going to rule the other way."

"And rule against the recommendations of my science staff?" I said.

"You must be joking. Seriously, I had no choice but to follow the recommendations of my technical staff...Now let's all relax for a few minutes and discuss how we are going to handle this dumping procedure."

"It should be done from the bridge where we can observe with the side cameras," said Mr. Barstow. "In fact, to facilitate this procedure, I must visit engineering."

"Very well, Mr. Barstow," I said.

"You may take care of that business." He left the bridge, and I turned my attention to the ladies. "I must also take care of some business, so I will see you back here in one hour."

CHAPTER 60

We all were gathering on the bridge where preparations for the dumping were taking place. Mr. Barstow entered data, and all of us were nervous.

"Who will do the dumping?" asked Ms. Dennison.

"I believe Ms. Cooper should, which is why I asked her to join us here," I said. "She normally operates the collection tanks, so she knows them best."

We all formed a circle as if engaged in some kind of ritual, and Ms. Cooper came forward and placed her hand above the control. We all looked around nervously.

"Now," I said, and she pressed the button. The dumping was complete, and no one called for a celebration. There was an awkward silence for a few moments.

"Does anyone have anything to say?" I asked. No one spoke.

"Let's get back to our stations," I said. Everybody left, and I sat down at the console.

"How are things at the helm, Mr. Blake?"

"Under control and sailing smoothly."

"Any expectations?"

"An uneventful journey home."

"Sounds kind of dull," I said. "I love dull."

Ms. Yates appeared at the doorway. "Captain, you are wanted on the phone."

"That's an interesting way to put it."

"Isn't that the way it operates?"

"I guess, Ms. Yates."

She handed me the remote unit, and I spoke. "This is Captain Jenner. Who am I speaking to?"

"It's Paul. I just wanted to see if this damn thing worked."

"Haven't you been receiving our download?"

"Yes, but I wanted to hear some direct communication."

"Well, you are hearing it now, but I have to get off the line. These phone bills are killing me."

"I'm sure as hell not paying them. Goodbye, Felix."

Ms. Grason appeared at the door and looked anxious to speak to me. "Captain. Since our party plans were interrupted the last time, I'm planning a celebration dinner this evening."

"Ms. Grason. I think that is an excellent idea, and I approve."

"Thank you, Captain. I've got to get busy. I'll see you at dinner."

She went on her way, and I turned my attention to the console. "That's certainly cause for celebration," said Mr. Blake. "You like our organized dinners, Mr. Blake?"

"They remind me of my mother's cooking."

"Your mother was a good cook?"

"The best."

"My mother was a psychologist."

"Captain, you have a scheduled meeting with Mr. Barstow in twenty minutes."

"Thanks, Mr. Blake. Tell Mr. Barstow to join me in the conference room."

I took the time to brew some coffee and relax. It was not easy to relax though, so I gave up trying and started pacing instead. Pacing turned out to be fun. Why hadn't I tried this before?

"Are you pacing, Captain?" said Barstow as he entered. "It's a new hobby."

"You sound more like yourself, Captain."

"Is that a good thing, Mr. Barstow?"

"That has yet to be determined. However, some things have been determined about our substance. We were not only being altered. We were being changed and added to. We are all evolving in unknown ways."

"Have you discovered any negative effects?"

"No, sir. I hope we don't, but we can't be sure that they won't turn up in the future."

"And the consensus of opinion is that we can't take the risk."

"That's right, sir."

"Then nothing more that needs to be said. In matters of science, the science officer has the last say. You do not have to justify anything to me. If there is nothing else, this meeting is adjourned."

"Yes, sir."

I returned to my duties on the bridge and resumed my pleasant banter with Mr. Blake. M. Clark joined us, and for two hours, we spoke of Earth adventures. Many amusing stories were told, and for the first time in a long time, I saw smiling faces. This was exactly what I needed to put myself in a better mood.

I left the bridge to catch a short nap before dinner. Sleep was not forthcoming, so I decided to do some reading. I chose a science-fiction novel and found the story quite fascinating. There wasn't anything far-fetched about this story at all.

It was getting to be dinnertime when I heard something outside the door. I opened the door to see that everyone was headed to the conference room. Apparently, Ms. Grason had changed the location of our feast. "Are you ready to party, Captain?" said Ms. Yates as she danced by with a tray of food. I stepped back into my room to make sure I was at my party best.

When I came back to the corridor, Ms. Dennison was standing there. "Hi, sailor," she said. "Want a date?"

"Do you need an escort?"

"Definitely."

"I would be delighted to accompany you to the ball." We

followed the group down the corridor when I stopped to look her over.

"You look lovely, Ms. Dennison."

"I was hoping you would think so, Captain."

"Now let's be on our way."

We continued down the hall, engaged in superficial evening talk. For fear of arriving first, we slowed our walk and sat for a few minutes at one of the observation windows.

"This promises to be the event of the season," I said. "I'm glad I could get tickets."

"They must have cost you a fortune."

"Yes, they did. Do you think she'll be there?"

"No. Not after the fool she made of herself last year."

"That's right. She may still be serving time."

"Oh, surely she's escaped by now."

"Ms. Dennison. What the hell are we talking about?"

"Damned if I know," she said as we looked at each other and laughed.

"Do you think we should be on our way?" I asked. "Yes, I'm sure it's time."

CHAPTER 61

We again walked in to a round of applause. "Speech!" yelled Mr. Decker from the back of the room.

"Not tonight, Mr. Decker."

Mr. Clark seated us and poured the wine. "They've done it again," said Ms. Dennison. "Who's done what?" I asked.

"The Rodgers twins appear to have prepared another beautiful dinner."

"I expect nothing less from my culinary chemists, Ms. Dennison."

Everyone seemed to be having a good time, even Mr. Barstow. Ms. Cooper was getting melancholic again, but she had Mr. Palasco's shoulder to cry on. Our entire engineering staff was busy talking shop and playing cards with a deck supplied by Ms. Yates.

I stood to propose a toast, and the room quieted. "Ladies and gentlemen, this dinner is to celebrate the end of our mission. And although we are still several months away from home, we have completed our obligations, and we can relax and celebrate. So enjoy the evening." We all took a drink and then settled down for a great meal.

"Is that all you have to say?" said Ms. Dennison.

"Believe me, that's all they wanted to hear. Now let's enjoy our meal."

"What are you going to do when you get back to Earth, Captain?"

"I want to spend some time on the beach."

"You mean Pebble Beach."

"You read my mind."

"I think I just want to sleep in."

"Somehow I can't imagine that."

"I've lost my passion for adventure."

"I can't believe that."

"Believe it, buster."

Mr. Clark came with our food. "I hope it's as good as it looks," said Ms. Dennison. She was quick to sample the feast.

"Well," I said. "How is it?"

"It's great."

"I expected nothing less."

"I think the Rodgers twins missed their calling."

"That may be true, Ms. Dennison, but they also make a very good bomb."

I excused myself and went to the bar for more wine. Ms. Kane was sitting at the bar.

"Captain," she said. "I hope you aren't angry about our decision to dump the substance."

"It had to be done. You had to do what you felt was right."

"You seemed angry."

"I suppose I was. I wanted to bring back the greatest discovery of all time."

"So did we. You don't know how much…"

"I'm beginning to find out, and I'm sorry I acted like such a jackass. I would like you to join me at my table."

We walked back to my table and sat down. "Ms. Dennison, I brought Ms. Kane over to join us because I owe you both an apology for the way I acted. I should have been more supportive and not such a devil's advocate."

"Captain," said Ms. Dennison. "We had a few fights about it also. It was the most difficult decision we have ever made."

Mr. Barstow approached the table, and I stood and greeted him with a handshake.

"I'm glad that I have you all together. I want to apologize to all of you for not supporting you in your decision regarding the substance. I had no right to second-guess your decisions."

We drank our wine and exchanged pleasantries, but we all knew something wasn't right. Maybe nothing would ever be right again. The positive feelings about discovering something great had been denied. I felt that we were going through the motions.

I wondered if everyone had that dark place that seemed to occupy more and more of my time. I'm not one to sit and ponder the ramifications of my actions, but questions kept creeping into my consciousness. The burden of responsibility was weighing me down. The contemplation of alternatives was depressing and made me wonder if my performance was affected by circumstances. My nightmares had turned to pleasant dreams, and reality had turned to a nightmare. Could these thoughts be the manifestation of the negative effects of the substance that had yet to be discovered?

The return home was not the easy journey I had hoped it would be. It became more difficult to maintain a cheerful demeanor. Paranoia was replacing confidence, and I felt I was losing control of my reasoning. How long would it take before the signs revealed themselves? Fear had replaced reasoning, and regardless of its cause, it was unacceptable. I began assigning more duties to Mr. Barstow.

The mood aboard ship was somber and reserved. Laughter had disappeared from the everyday activities, and a strictly-business attitude prevailed. The thought of the possible ramifications of our medical condition must have weighed heavily on everyone's mind.

What if we did not know of the infection? Would we all be happy and looking forward to the return home? Maybe the blanket of depression was more the product of our own paranoia. I remembered my father's advice about not worrying over things that you can't control. That would be hard to do, but I had a esponsibility to the crew. I scheduled more parties, and it did seem to improve the general mood of the crew. From that point on, I made it my prime concern to keep a happy crew.

My efforts did not go unnoticed. Ms. Kane and Ms. Dennison both expressed their appreciation, Ms. Dennison adding that the wine at the parties didn't hurt. I knew I was on the right track after a remark like that.

Things were improving, and research continued. Mr. Barstow was learning more about the alien's quantum weapon, our new communication system continued to function perfectly, and I even got back on the handball court with Ms. Yates. Even the Rodgers twins continued to make those incredible meals, which, I felt, was the single most important element to the improvement of morale.

The next few months seemed to go as well as could be expected. As we approached the docking date, a certain tension in the air increased but that was also to be expected.

It would take a long time before anyone on the USS Mobius would feel like normal, and since all of us were infected, we would be the subjects of much scientific examination. Our lives would never be the same again. As the days passed, my thoughts returned to the encounter with the aliens—what potential threat they may be had yet to be determined. We may have lost them, but we would have a confrontation…one day.

The End

ABOUT THE AUTHOR

As an author of articles for Popular Electronics magazine, www.electroschematics.com, and Nuts & Volts Magazine and robotics articles for Servo Magazine and quantum physics articles for an online publisher, www.electricalfun.com—Terence decided to use his technical knowledge and applied it to a novel. He has always been a fan of science fiction, and Outer Domain had been floating around in his mind for some time.

Short stories of his, with a science-fiction or horror genre, have been published in books like Endlands, Volume 2 and Of Sun and Sand and online magazines like Tigershark magazine, www.kindofahurricanepress.com, and www.kalkion.com.

One of the most popular books on the subject of electronic music is Terence's Sound Synthesis: Analog and Digital Techniques. Poems like "I Am" have been published in a book, Forever Spoken, a 2007 book from the International Library of Poetry. His academic credits include being chief engineer for the electronic music department at New York University and a teacher at the Mannes College of Music, the New School for Social Research, and FreeSpace Alternate University—all in New York City.